W9-CLR-184

HOW INSURGENCIES END

Ben Connable and Martin C. Libicki

Prepared for the Marine Corps Intelligence Activity

Approved for public release; distribution unlimited

NATIONAL DEFENSE RESEARCH INSTITUTE

The research described in this report was prepared for the Marine Corps Intelligence Activity. The research was conducted in the National Defense Research Institute, a federally funded research and development center sponsored by the Office of the Secretary of Defense, the Joint Staff, the Unified Combatant Commands, the Department of the Navy, the Marine Corps, the defense agencies, and the defense Intelligence Community under Contract W74V8H-06-C-0002.

Library of Congress Cataloging-in-Publication Data

Connable, Ben.
　How insurgencies end / Ben Connable, Martin C. Libicki.
　　p. cm.
　Includes bibliographical references.
　ISBN 978-0-8330-4952-0 (pbk. : alk. paper)
　1. Insurgency--Case studies. 2. Counterinsurgency--Case studies. I. Libicki, Martin C. II. Title.

　JC328.5.C665 2010
　355.02'18--dc22

2010014522

The RAND Corporation is a nonprofit research organization providing objective analysis and effective solutions that address the challenges facing the public and private sectors around the world. RAND's publications do not necessarily reflect the opinions of its research clients and sponsors.

RAND® is a registered trademark.

Cover design by Carol Earnest

© Copyright 2010 RAND Corporation

Permission is given to duplicate this document for personal use only, as long as it is unaltered and complete. Copies may not be duplicated for commercial purposes. Unauthorized posting of RAND documents to a non-RAND Web site is prohibited. RAND documents are protected under copyright law. For information on reprint and linking permissions, please visit the RAND permissions page (http://www.rand.org/publications/permissions.html).

Published 2010 by the RAND Corporation
1776 Main Street, P.O. Box 2138, Santa Monica, CA 90407-2138
1200 South Hayes Street, Arlington, VA 22202-5050
4570 Fifth Avenue, Suite 600, Pittsburgh, PA 15213-2665
RAND URL: http://www.rand.org/
To order RAND documents or to obtain additional information, contact
Distribution Services: Telephone: (310) 451-7002;
Fax: (310) 451-6915; Email: order@rand.org

Preface

Insurgencies have dominated the focus of the U.S. military for the past seven years, but they have a much longer history than that and are likely to figure prominently in future U.S. military operations. Thus, the general characteristics of insurgencies and, more important, how they end are of great interest to U.S. policymakers.

This study constitutes the unclassified portion of a two-part study that examines insurgencies in great detail. The research documented in this monograph focuses on insurgency endings generally. Its findings are based on a quantitative examination of 89 cases. James Bruce is the overall project manager.

This research was sponsored by the U.S. Marine Corps Intelligence Activity (MCIA) and conducted within the Intelligence Policy Center (IPC) of the RAND National Defense Research Institute, a federally funded research and development center sponsored by the Office of the Secretary of Defense, the Joint Staff, the Unified Combatant Commands, the Navy, the Marine Corps, the defense agencies, and the defense Intelligence Community.

For more information on RAND's Intelligence Policy Center, contact the Director, John Parachini. He can be reached by email at John_Parachini@rand.org; by phone at 703-413-1100, extension 5579; or by mail at the RAND Corporation, 1200 South Hayes Street, Arlington, Virginia 22202-5050. More information about RAND is available at www.rand.org.

Contents

Figures

Tables

Summary

Purpose and Approach

This monograph describes and explains insurgency endings in order to inform policy and to guide strategic and operational analysis.

Our methodological approach had two components. One was a review of insurgency and counterinsurgency (COIN) literature. The second was a quantitative and qualitative analysis of 89 insurgency case studies.

"How Insurgencies End" has produced several findings, some of which reinforce or explain conventional wisdom regarding insurgency and COIN. Others present a new range of dilemmas and opportunities to policymakers and planners. We derived additional findings primarily from the quantitative research. These findings reveal some useful insights into the relative success or failure of various methods employed by each side as they apply to insurgency endings. A few of these additional findings describe the impact of existing operational and environmental factors on COIN operations, thereby informing policy decisionmaking.

Finally, we describe a small set of key indicators drawn from the results of the study. RAND identifies key indicators as those found to be broadly applicable across a range of operational environments. Other interesting but less broadly applicable indicators are mentioned in the body text. The reader will find that conventional wisdom regarding tipping-point indicators in COIN is strongly reinforced by the research conducted for *How Insurgencies End*.

Generalized findings derived from historical case studies should not be taken as prescriptions for upcoming or ongoing operations. For example, while we found that insurgencies last about ten years, we do not suggest that any specific insurgency will last ten years. Predicting specific outcomes from general assumptions is a common logical or, by some scientific definitions, an ecological fallacy. Quantitative findings alone cannot—and should not—shape COIN campaign planning. Further, these findings are correlative and not necessarily causative. In other words, addressing only one factor, such as the availability of sanctuary or external support, will not necessarily create a tipping point that will mark the beginning of the end of the insurgency.

Key Findings

Modern insurgencies last approximately ten years, and the government's chances of winning may increase slightly over time. Quantitative analysis of the 89 cases selected for this study shows that the median length of an insurgency is ten years. An insurgency that hits a clear tipping point at or just before ten years typically tails out gradually to end state at 16 years. This finding squares with conventional wisdom, as well as existing conclusions published in a range of COIN literature. Furthermore, although the statistical data show only a weak relationship between time and outcome, the longer an insurgency lasts, the more likely the government is to win.

Based on this finding, a counterinsurgent might assume that a campaign will typically play out along a ten-year arc. However, this *does not* mean that every campaign will last ten years. At best, this finding plants a marker for basic planning assumptions and belies the notion that insurgents can win simply by surviving. Taking into account all other factors, governments executing COIN campaigns approaching the apex of this ten-year arc should not assume that insurgent perseverance foreshadows an insurgency's eventual success.

The tailing nature of government victories—where the violence ebbs slowly rather than rapidly, as in typical insurgent victories—reflects the concept of the tipping point, explained in the body text.

An insurgency could effectively be over without either side realizing that it had won or lost for several years. As evinced in several of the 89 cases, "defeated" insurgencies can splinter into smaller, more-violent terrorist organizations or hibernate with the intent of reigniting hostilities when conditions present themselves. Formation of the Real Irish Republican Army (RIRA) in the wake of the Good Friday Agreement and the reemergence of the Peruvian Shining Path guerrillas illustrate these two trends.

Withdrawal of state sponsorship cripples an insurgency and typically leads to its defeat. Inconsistent or impartial support to either side generally presages defeat. State sponsorship is simply defined as either direct or indirect support provided to an insurgency by a nation-state (e.g., the United States supporting the Afghan mujahideen against the former Soviet Union during the 1980s) or by one nation-state to another. Sponsors provide direct support in the form of military intervention—through kinetic strikes, deployed troops, or deployed trainers—or indirect financing and equipping. Insurgencies that we studied that benefitted from state sponsorship statistically won at a 2:1 ratio out of decided cases. When that sponsorship was wholly withdrawn (e.g., Greece, 1945–1949), the victory ratio for the insurgent fell to 1:4 (also of decided and not mixed or ongoing cases). In other words, loss of state sponsorship correlates with a tipping point. Loss of sponsorship frequently correlates with loss of sanctuary, a critical requirement for insurgents. The effect of the withdrawal of state sponsorship is one of the most statistically robust findings in this study. We will show the crippling effects of inconsistent support to insurgencies over time in the Angola and South Thailand cases. In the Angola case, consistent support to the Movimento Popular de Libertação de Angola (MPLA, or Popular Movement for the Liberation of Angola) insurgents allowed that organization to eclipse the União Nacional para a Independência Total de Angola (UNITA, or National Union for the Total Independence of Angola) insurgent group that received support from over a dozen countries but dedicated support from none. Inconsistent or untimely termination of support for governments embroiled in a COIN fight can be equally crippling, as evinced in the Vietnam and Cuba cases, among others.

Anocracies (pseudodemocracies) do not often succeed against insurgencies and are rarely successful in fully democratizing. Fifteen of the 89 cases we studied could be described as *anocracies*, or democracies in name only. Anocracy is a particularly weak form of government in that it is good at neither democracy nor autocracy: It gains little benefit from reform and must refrain from using effective repressive tactics in order to retain the façade of democracy. Anocracies have a particularly poor record at countering insurgency, winning about 15 percent of all contests (1:7, with eight ongoing or mixed outcomes). Lessons from the one case of successful democratization we identified—Croatia—are both debatable and not necessarily transferrable to other conflicts. Democratizing an anocracy in the midst of an insurgency is an unappealing but not necessarily impossible venture.

Key Indicators

Desertions, Defections, and Infiltrations

The rates at which these phenomena occur, as well as changes in these rates, often indicate significant trends and, occasionally, tipping points. *Desertion* occurs when an insurgent or government agent (soldier, officer, or civil servant) flees control of the parent organization without necessarily defecting to the other side. *Defections* occur when a group or subset of belligerents changes sides, willingly entering the opposing camp. *Infiltrations* are the introduction of friendly elements into the opposing belligerent's organization or the covert "turning" of belligerents to act as informants.

In analyzing the trends associated with defection and desertion rates in particular, it is important not to confuse definitions. Although deserters may also defect, in many cases, they may simply flee. Assuming that accurate information is available, defection and desertion statistics should be tracked separately and closely analyzed for relative value. If deserters do not also defect to the other side, this may indicate disillusionment with both sides, or it could reflect a range of other factors (e.g., the insurgent returns home for seasonal harvesting with the intent of eventually rejoining the insurgency). It is also important to analyze who is deserting. For example, officer desertions may have greater value than foot-soldier desertions. Simple desertions should not necessarily be taken as a significant indicator of success for one side or the other.[1]

Information and Reporting

Qualitative analysis of the 89 cases substantiates conventional wisdom in that civilians' willingness to report on insurgent activity to the government usually reflects the success of government security and pacification programs.[2] Conversely, a lack of reporting correlates with a lack of progress. A change in the frequency and quality of reporting in either direction may mark a shift in momentum, particularly at the tactical level. Although it is difficult to quantify this kind of reporting, some counterinsurgents have tracked the number and quality of tip-line calls as a metric of success. Tip-line metrics are of dubious value, especially considering the number of ways in which tip-line calls can be spoofed.[3] In the best case, all methods of reporting (telephonic, face-to-face, written) would be tracked.

---

[1]   A declassified intelligence assessment prepared for the director of the Office of Current Intelligence of the Central Intelligence Agency in 1966 (Adams, 1966) aptly addresses the necessity to delink *defection* and *desertion* to prevent analytic fallacy.

[2]   Insurgents also rely heavily on "street intelligence" and can similarly track their progress by the rate or flow of information.

[3]   Urban insurgent Carlos Marighella suggests spoofing tip-lines and mail to call in false bomb reports and other acts of terrorism to disrupt government COIN operations.

Additional Findings

Complex Insurgency, Complex and Protracted Ending

Insurgencies with more than two clear parties involved have longer, more-violent, and more-complex endings. Afghanistan is a case in point, with at least seven major ethnic groups, several insurgent organizations, Iran, Pakistan, India, the United States, the North Atlantic Treaty Organization (NATO), and various other actors all claiming a stake in the outcome of the conflict. While not all parties must be satisfied to bring an end to the immediate conflict, the dissatisfaction of one or more parties will probably complicate the ending and may allow the insurgency to smolder and eventually reignite.

Governments Outlast Insurgents

Contrary to conventional wisdom, insurgents do not win by trying to simply outlast the government. In fact, over the long run, governments tend to win more often than not. This finding seems to belie the notion that the factor of time always runs against the counterinsurgent.

Governments Are Better Off Without Support

Governments benefit from direct support but tend to lose more frequently when provided indirect support; *they do slightly better with no external support at all*. Once support is given, it almost always creates a dependency on the external sponsor. We contrast the case of U.S. support to South Vietnam with Chinese (and Soviet) support to the north.

Insurgency Is Suited to Hierarchies and Rural Terrain

Unified hierarchies do better at insurgency than do fragmented networks. Most insurgencies consist of a hybrid of these two models, but urban insurgencies tend to be more networked than their rural counterparts. This finding is closely linked to the finding that insurgencies rarely succeed in middle-income and urbanized countries. Insurgency is an endeavor best practiced in rural, or a mix of rural and urban, terrain.

Terrorism Often Backfires

Broad terror campaigns by insurgents correlate with insurgent defeat, but selective terror attacks that do not kill innocent civilians correlate with a marked insurgent advantage. In other words, those insurgent groups that were able to restrict their use of terrorism by minimizing civilian—vice government—casualties were more likely to win than those that did not. Qualitative analysis shows that the use of indiscriminate terror often is a sign of overconfidence or, conversely, of weakness.

Weak Insurgents Can Win

Insurgents do not need to be militarily strong to win, and, in fact, military strength can backfire if the threat of insurgent military victory galvanizes government security forces. In cases of long-running insurgencies, like those in Colombia and Sri Lanka, the government was able to reinvigorate COIN efforts even in the face of powerful insurgent cadre (in this case, the Fuerzas Armadas Revolucionarias de Colombia [FARC, or Revolutionary Armed Forces of Colombia] and the Tamil Tigers of Eelam).

Sanctuary Is Vital to Insurgencies

Availability of sanctuary directly correlates with an improved likelihood of insurgent victory, but only if it is provided voluntarily. Insurgencies rarely survive or succeed without some kind of sanctuary. Internal sanctuary is also very valuable, perhaps as valuable as voluntary, external sanctuary.

Conclusion

Full-blown insurgencies are messy affairs. External sponsors sometimes back winning causes but rarely emerge with a clear victory. Not one of our 89 cases provided an example that could be equated to an unambiguous conventional success like that of the Allies in World War II. Recent U.S. experience in COIN has been especially tangled. Vietnam speaks for itself, as do Iraq and Afghanistan. This kind of mixed outcome is all but inevitable. However, ways exist to mitigate negative consequences. It is possible to shape insurgency endings with sufficient forethought, strategic flexibility, and sustained willpower.

When viewed holistically, the data set for this study revealed a few trends consistent to all insurgency endings. In some cases, these trends reinforce conventional wisdom and lend credibility to COIN advocates shaping U.S. policies on Iraq and Afghanistan. At the same time, they should give pause to those advocating for alternative approaches—specifically, the use of indirect tactics and strategies that separate the counterinsurgents from the population. In nearly all cases we studied, only the direct and consistent application of basic COIN methodology promulgated by David Galula (1964 [2006]), David Kilcullen (2009), Thomas X. Hammes (2006), GEN David Petraeus, Gen. James Mattis, and others leads to favorable endings. Failure to heed the past 50 years of expert opinion on the subject almost guarantees an undesirable, and possibly a disastrous, end.

Acknowledgments

James Bruce oversaw the project. Anthony Butera, Peter Chalk, Sara A. Daly, Brian A. Jackson, Seth G. Jones, William Rosenau, Paraag Shukla, and Anna-Marie Vilamovska conducted the quantitative examination of 89 insurgencies. Jerry M. Sollinger edited the final draft and provided many helpful comments and suggestions along the way. Stephen T. Hosmer, Daniel Byman, Jasen J. Castillo, Katharine Watkins Webb, John Gordon, and Christopher Paul all offered mentorship and critical guidance during the quality-assurance process. Ben Wise provided insight into our mathematical modeling processes.

Brian A. Jackson, Peter Chalk, Sara Daly, Anthony Butera, John Gordon, and Austin Long contributed detailed case studies during the research phase of this project. We liberally referred to these studies while writing the final draft of this report but were unable to include them in full. Each author is in large part responsible for shaping the qualitative elements of our research.

We would like to give our special thanks to (as of mid-2009 Maj. Gen.) John F. Kelly, U.S. Marine Corps (USMC), for his encouragement and comments on the first draft of this monograph, and to Lt. Col. Drew E. Cukor, USMC, for his valuable insight and contributions.

Abbreviations

ADF	Arab Deterrent Force
ALF	Afar Liberation Front
AMS	Association of Muslim Scholars
ANC	African National Congress
AO	area of operations
AP3	Afghan Public Protection Program
AQI	al Qaeda in Iraq
AQIM	al Qaeda in the Islamic Maghreb
ARVN	Army of the Republic of Vietnam
BRN	Barisan Revolusi Nasional, or National Revolutionary Front
C2	command and control
CAP	Combined Action Program
CAT	civil-action team
CDF	civil-defense force
CIA	Central Intelligence Agency
COIN	counterinsurgency

CORDS	civil operations and revolutionary development support
DRV	Democratic Republic of Vietnam
DSE	Democratic Army of Greece
ECOMOG	Economic Community Military Observation Group
ETA	Euskadi Ta Askatasuna, or Basque Homeland and Freedom
FARC	Fuerzas Armadas Revolucionarias de Colombia, or Revolutionary Armed Forces of Colombia
FATA	federally administered tribal area
FIS	Front Islamique du Salut, or Islamic Salvation Front
FLN	Front de Libération Nationale, or National Liberation Front
FM	field manual
FNLA	Frente Nacional de Libertação de Angola, or National Liberation Front of Angola
FNLC	Front for the National Liberation of the Congo
FPS	Facilities Protection Service
FSLN	Frente Sandinista de Liberación Nacional, or Sandinista National Liberation Front
GIA	Groupe Islamique Armé, or Armed Islamic Group
GSPC	Groupe Salafiste pour la Prédication et le Combat, or Salafist Group for Preaching and Combat
GVN	government of Vietnam
GWOT	global war on terror

Hamas	Harakat al-Muqāwamat al-Islāmiyyah, or Islamic Resistance Movement
Huk	Hukbalahap
IRA	Irish Republican Army
ISI	Inter-Services Intelligence
JVP	Janatha Vimukthi Peramuna, or People's Liberation Front
KIA	killed in action
KKE	Kommounistikó Kómma Elládas, or Communist Party of Greece
LRA	Lord's Resistance Army
LTTE	Liberation Tigers of Tamil Eelam
MCIA	U.S. Marine Corps Intelligence Activity
MILF	Moro Islamic Liberation Front
MNLF	Moro National Liberation Front
MPLA	Movimento Popular de Libertação de Angola, or Popular Movement for the Liberation of Angola
NATO	North Atlantic Treaty Organization
NPA	New People's Army
NVA	North Vietnamese Army
ONR	Office of Naval Research
PIRA	Provisional Irish Republican Army
PKK	Partiya Karkerên Kurdistan, or Kurdistan Workers' Party
PLO	Palestine Liberation Organization

PRC	People's Republic of China
PULO	Patani United Liberation Organization
RENAMO	Resistência Nacional Moçambicana, or Mozambican National Resistance
RIRA	Real Irish Republican Army
SAM	surface-to-air missile
SAODAP	Special Action Office for Drug Abuse Prevention
SoI	Sons of Iraq
SPLA	Sudan People's Liberation Army
TEA	Tagmata Ethnofylakha Amynhs, or Greek National Guard Defense Battalions
UNITA	União Nacional para a Independência Total de Angola, or National Union for the Total Independence of Angola
USMC	U.S. Marine Corps
VC	Vietcong
USSR	Union of Soviet Socialist Republics
WTC	World Trade Center
WWW	World Wide Web

CHAPTER ONE

Introduction

Insurgency has been and will continue to be a consistent feature of the security environment. Within the coming decades, U.S. policymakers and strategic planners will almost certainly face dilemmas and decisions similar to those faced in the days and months leading up to Operations Enduring Freedom and Iraqi Freedom. To enable better planning for these likely challenges, it is critical to understand how insurgencies end.

This understanding will help answer the most-important questions posed at the leading edge of the national-security decisionmaking process: Is the prospective operation viable? Is it worth the anticipated risk to international prestige and treasure? Do conditions on the ground seem to favor a successful counterinsurgency (COIN) campaign, or do they suggest failure? What are the likely long-term costs associated with securing the populace to achieve the desired goal?

If the operation is deemed practicable, knowledge of insurgency endings can inform the design of the COIN campaign and help mitigate the kind of false expectations that undermined the arc of the conflicts in Iraq and Afghanistan. It provides a realistic planning framework for both policymakers and strategists. An appreciation for how insurgencies end provides planners with a valuable instrument with which to help manage or, possibly, reduce the suggested ten- to 16-year timeline of the typical insurgency.

Purpose of This Monograph

In addition to explaining how and why insurgencies end, this monograph has three ancillary but supporting objectives. First, we attempt to describe and qualify conventional wisdom regarding insurgency and COIN. Second, based on a quantitative analysis of 89 insurgencies, we describe the common trends of those that succeed and those that fail. Third, we provide transferrable *end-state* indicators for intelligence professionals. If properly evaluated, these indicators can help counterinsurgents recognize, or create, a *tipping point*. The concept of the tipping point is essential to unlocking the central question of this study: How do insurgencies end?

Malcolm Gladwell (2000) described the concept of the tipping point in his best-selling book. Although the term is fairly self-explanatory—it is the point at which events take a crucial turn toward the final outcome—the identification of the factors that generate a tipping point are often elusive. Further, it is commonly very difficult to recognize a tipping point until long after it has passed. Gladwell reports that most tipping points occur unbeknownst to even close observers and that these observers often draw erroneous conclusions from paradigm shifts in conditions or behavior.

The qualitative elements of this project focus in part on identifying and describing the tipping points in a selected set of COIN campaigns. These points mark the beginning of the end of the insurgency. What factors led to this point? What event, action, or lack of action was most significant in generating a tipping point that ended the insurgency? We also examine and describe the events subsequent to the tipping point. By overlaying these qualitative assessments on quantitative analysis of insurgency trends, useful indicators can be described.

Not every insurgency had, or has, a tipping point. Many insurgencies end in drawn-out negotiated settlements, some of which are inconclusive from the perspective of both the insurgents and the government. Some cases with seemingly clear-cut endings had, at one point, tipped against the insurgency only to tip back again years later when the insurgents emerged from hibernation or external sanctuary. In this respect, a tipping point does not signal an irreversible event.

Instead, it is used as a descriptive device to explain historical cases and as a marker for intelligence analysis. We provide a more extensive explanation of the concept using an example in the next chapter.

Because so many unique variables define insurgencies, including local culture, terrain, economy, and government, to name only a few, we found that only a small set of indicators is suitable for generalization. Attempting to draw generalized lessons from insurgencies is, at best, an inexact science and, at worst, informed speculation. Broad surveys can devolve into exercises in simple reiteration, while narrow case studies offer few universal truths. Recognized COIN expert David Kilcullen sums up the various pitfalls this way:

> [There] is no standard set of metrics, benchmarks, or operational techniques that apply to all insurgencies or remain valid for any single insurgency throughout its life cycle. And there are no fixed "laws" of counterinsurgency. . . . (Kilcullen, 2009, p. 183)

However, many distinguished experts (including Kilcullen) have distilled a few general lessons. Seminal works by Mao Tse-tung, David Galula, and many others all address insurgent and counterinsurgent strategies that can arguably prove useful in a variety of operational environments. A considerable body of existing literature speaks to this conventional wisdom and to the questions posed in this research, a brief review of which is provided later in the introduction. These lessons portray a loose outline of conventional wisdom on insurgency and COIN.

We do not intend a simple restatement of these existing hypotheses or a rehashing of conventional wisdom. Instead, we use a detailed examination of quantitative and qualitative data to explain, justify, or refute convention. We believe that an in-depth study of such a sizable case sampling would also be likely to produce some unexpected results; in a few cases, this proved true. Some data will necessarily prove inconclusive or quantitatively insignificant.

A Note on Contemporary Threats and Operations

This monograph reached publication when the United States was simultaneously committed to two major COIN operations. The monograph will necessarily be judged in relation to these commitments: It will be expected to provide contemporaneous, relevant findings. Indeed, the U.S. Marine Corps Intelligence Activity (MCIA) initially commissioned the umbrella project for *How Insurgencies End* in order to gain perspective on Marine COIN operations in Al Anbar province in Iraq.

As operations in Anbar province wound down, we had to choose between two presentation approaches. The first would directly compare each finding with some of the broader issues relating to Operations Enduring Freedom and Iraqi Freedom. Following this approach, the final report would describe the immediately relevant data while trimming sections deemed inapplicable or distracting. Such a targeted study, however, would be less germane to other ongoing and prospective operations. It would inevitably lose value over time.

Instead, we chose to present broader findings with the intent of contributing to the body of COIN literature. In this way, policymakers and strategists can draw their own conclusions from an analysis of the historical record. Afghanistan (2001–present) and Iraq (2003–present) are counted among the 89 cases, but this monograph only selectively refers to ongoing U.S. COIN operations to support quantitative analysis and to help explain quantitative trends.

Additionally, we chose not to focus on global Islamist or other religious threats, except as they relate to individual insurgency cases. We believed that the study would quickly bog down in an examination of conventional wisdom on the subject of global terrorism, which, in any case, only coincidentally applies to the study of insurgency and COIN.

Research Approach

Our research approach involved two steps. First, we identified and reviewed the literature on insurgencies and COIN operations. Among other things, this review provided us a basis for identifying what we call the conventional wisdom regarding insurgencies and how they end. We then decided on a case-study approach as a way of testing the validity of that wisdom and identifying additional information relevant to the ending of insurgencies. But to do that, we had to determine which cases we would analyze. Our start point was the data set of insurgencies from James Fearon and David Laitin (2003a), which included 127 insurgencies. We winnowed that number down, and we added some of our own selection (e.g., insurgencies that postdate Fearon and Laitin's work) arriving at a set of 89 that we used for our analysis. The rationale for adding and subtracting cases is explained in Appendix A; Appendix D contains a list of the insurgencies we dropped from consideration.

Having identified the insurgencies we would study, we then approached them in two ways: one quantitative and one qualitative. In the qualitative phase, we developed a simple taxonomy of endings: government wins, insurgents win, mixed, and still ongoing. Every insurgency was assigned to one of the four outcomes based on our assessment. In 28 cases, the government won. In 26 cases, we judge the insurgents to have prevailed. In 19 cases, we view the outcome as mixed in that neither side achieved all it wanted. Sixteen insurgencies have yet to conclude.

We then parceled out the 89 insurgencies to experienced RAND analysts and research assistants, most of whom had enough knowledge of the region or insurgency to reach conclusions on their character based on earlier research. We asked them to identify variables pertaining to the insurgency, the government, and the country. This exercise was performed twice. The first survey was carried out the spring of 2006 with an emphasis on the factors that led to the government's winning and losing. The second survey was carried out in the autumn of 2006 with more of an emphasis on how the insurgencies were brought to an end. The goal was to determine what variables were associated with a given type of outcome.

The data used in the development of the quantitative portion of the study were collected in 2006. This means that the statistics driving the study stem from that year. While the qualitative analysis of the data has been updated, the numbers themselves will show some anomalies when compared to current events. For example, we rated the Taliban insurgency in 2006 as marginally competent. That would certainly not be true today, and one could retroactively argue it was not true in early 2006 as the Taliban prepared for its unexpected surge.

The qualitative portion of the approach consisted of incorporating examples drawn from the 89 case studies with the quantitative data to provide descriptive grounding to the statistical analysis. This combination should produce reasonably robust and, in some cases, more–broadly applicable findings.

Graphs Used in This Monograph

Graphically portraying a qualitative assessment presents several challenges. It would be imprudent to attempt to depict the ebb and flow of an insurgency using numerical scale, to accurately plot movement along x- and y-axes, or to show direct relationships to any degree of certitude. Instead, this monograph relies on a series of notional graphs interspersed throughout the body text to depict the ebb and flow of insurgencies as they play out to end state. These graphs are intended to help the reader visualize insurgency endings. They are derived from qualitative and quantitative research, showing general trends over the course of an insurgency and, more specifically, end state in relation to the overall "arc" of the conflict. Plots on the graphs are not tightly linked to x- and y-axis increments. We have modeled these notional depictions on a graph published by COIN expert David Galula (1964 [2006], p. 61, fig. 3) in *Counterinsurgency Warfare: Theory and Practice*.

Conventional Wisdom

We use this term in a complimentary rather than pejorative sense, in that it helped us to isolate what we consider to be the authorities in the fields of insurgency and COIN. Hundreds, if not thousands, of authors have published works on insurgency theory, practice, and (predominantly) specific case studies. The sheer volume of literature required us to narrow our focus to include only recognized experts as we attempted to frame conventional wisdom. Because the term *conventional wisdom* denotes inherent ambiguity, we assume a bit of leeway in the selection process; we did not limit ourselves to existing, official reading lists. And, as conventional wisdom is also prevailing wisdom, most of the works we selected were published after 1945 (with Mao Tse-tung's work being an obvious exception). We focused on identifying philosopher/practitioners, or those authors who had both studied *and* practiced either insurgency or COIN. However, we include notable scholarly works on insurgency and civil violence in order to balance the experiential writings with objective rigor.

The resulting field of experts contains names, theories, and narratives that we believe have shaped conventional wisdom on insurgency and COIN. By referring to these works throughout the monograph rather than in a single chapter, we hope to assist the reader in contrasting each of our findings with some of the assumptions that have advised recent operational planning. The following is, in essence, a very brief literature review that sets the stage for later citation.

With a few notable exceptions (including al Qaeda in Iraq, or AQI), nearly all contemporary insurgency theory—rural communist, *foco*, and urban—is rooted in Mao Tse-tung's (1961 [2000]) *On Guerrilla Warfare*.[1] Mao laid the foundation for the rural communist insurgencies of Ho Chi Minh and Ernesto "Che" Guevara, while inspiring and shaping nearly every other insurgency since the early 1960s.[2] General Võ Nguyên Giáp (1961 [2000]) brought to print the insurgent philosophy and history of Ho Chi Minh in *People's War, People's Army*, a work that builds on Mao's *On Guerrilla Warfare* and served to inspire contemporary and later insurgencies. For his part, Guevara split with

---

[1] There are several different translations and transliterations of this title. We refer to it as cited in our bibliography.

[2] Although we generally limited ourselves to citing *On Guerrilla Warfare*, Mao's insurgent philosophy is recorded and deciphered in a number of excellent volumes. *Mao Tse-Tung in Opposition, 1927-1935* (Rue, 1966) and *Basic Tactics* (Mao, 1966) are somewhat obscure, but additional translations of Mao's own work can be found in a multivolume series, *Selected Works* (Mao, 1954–1962).

Mao in *Guerrilla Warfare*, espousing the fanciful *foco* modification to Chinese communist rural insurgency theory.[3] Régis Debray supplemented Guevara's *Guerrilla Warfare* with *Revolution in the Revolution?*, a somewhat more philosophical work that also fed the disastrous notion of the rural *foco*. It was not difficult to select the controversial Carlos Marighella to represent urban philosophy. Marighella's (2008) *Minimanual of the Urban Guerrilla* builds on Mao and Che to a point and then diverges, all but abandoning the themes of rural land reform and the long war in favor of the kind of quick, explosive urban campaigns that shook Latin America in the late 1960s and early 1970s.[4] We also cite Abraham Guillén. While Guillén preached urban insurgency, he also espoused the kind of hybrid rural/urban insurgency that has emerged in several 21st-century conflicts.[5] In this way, Guillén may be the most visionary of our insurgent theorists. Finally, Robert Taber (1965) presents an interesting, if rather subjective view of insurgent philosophy and practice in *War of the Flea*.[6]

David Galula (1964 [2006]) leads the list of counterinsurgents with *Counterinsurgency Warfare*, arguably the most recognized and influential book on the subject. Galula was an experienced practitioner who wrote with brilliant economy. Nearly all COIN philosophies stem from or refer to Galula either directly or indirectly, although many

---

[3] Which itself was derived from Marxist/Leninist theory.

[4] *Minimanual* is a tactical, crib-notes distillation of his more nuanced understanding of insurgency theory, and it represents a rather radical shift in perspective. In the anthology *For the Liberation of Brazil*, Marighella (1971, pp. 47, 179) states that "the decisive struggle will be in the rural area—the strategic area—and not the tactical area (i.e. the city)" and that "guerrilla warfare is not the right technique for urban areas." This earlier work is bogged down with boilerplate Marxist rhetoric, but it also offers some parallel analysis to Guillén's (1973) *Philosophy of the Urban Guerrilla*. The British COIN manual states, "The *Minimanual of the Urban Guerrilla* was to aspiring urban insurgents in the 1970s what Mao Tse-Tung's *Protracted War* (a.k.a. *On Warfare*) had been to earlier generations of rural revolutionaries, and for much the same reasons" (UK Ministry of Defence, 2001, p. A-1-E-1).

[5] Guillén offers a stinging critique of Marighella, Guevara, and others in the second-to-last chapter of *Philosophy of the Urban Guerrilla*. We recommend reading Guillén only after reading Mao, Che, and Marighella.

[6] Notably absent from this list are modern Islamic philosopher/practitioners, such as Osama bin Laden. In an effort to separate terrorism from insurgency, we chose to omit them.

of our philosopher/practitioners reached similar conclusions independently. We will cite him extensively. Galula is followed closely by David Kilcullen, whom one might describe as the "GWOT [global war on terror] Galula." Kilcullen is best known for his informally published *Twenty-Eight Articles* (2006) on COIN, but he also authored *Countering Global Insurgency* (2004) and, more recently, *The Accidental Guerrilla* (2009). Kilcullen has helped shape conventional wisdom not only through his writing but also through his association with the "surge" in Iraq and his work with noted COIN practitioner GEN David Petraeus (U.S. Army). John J. McCuen is a (now retired) U.S. Army colonel with experience in several Southeast Asian COIN operations and the author of *The Art of Counter-Revolutionary War* (1966). McCuen's book, obscure but occasionally cited by experts, is an overlooked resource.

Although Bernard B. Fall was a correspondent and historian rather than a military practitioner, we felt it appropriate to include *Street Without Joy* (1964) as an experiential title. Fall's firsthand narrative and his musings on the future of "revolutionary war" have influenced many of today's COIN experts, including Kilcullen.[7] Richard Clutterbuck, an experienced field hand turned professor, contributed to the body of literature with *Guerrillas and Terrorists* (1977) and *The Long, Long War: Counterinsurgency in Malaya and Vietnam* (1966). Jeffrey Record served as an assistant province adviser in Vietnam and authored the firmly worded *Beating Goliath* (2007). Retired Army officer John Nagl has produced a series of products on COIN, most notably *Learning to Eat Soup with a Knife* (2005). Nagl's title references T. E. Lawrence, whose works we used only as deep reference material.[8] Thomas X. Hammes rounds out this portion of our list.[9] Hammes'

---

[7] Kilcullen references Fall twice in *Accidental Guerrilla*.

[8] Although Lawrence is widely quoted in COIN circles, he was an insurgent adviser. While Lawrence has influenced conventional wisdom, we felt that many of his aphorisms were too narrow (or too general), too dated, and too far out of context to apply to our findings. We could have cited him as an insurgent, but we found little evidence that Lawrence's writings influenced modern insurgents to any great degree.

[9] We also decided not to cite Colonel C. E. Caldwell's voluminous *Small Wars: Their Principles and Practice* (1906 [1996]). Caldwell's book is no longer in the mainstream, and many of his observations are specifically tactical or exceedingly dated: The second edition of *Small*

The Sling and the Stone (2006) is on the required reading list at many military educational institutions. Hammes is one of the more concise and prescriptive authors, and we found that he often summarized conventional wisdom in a very digestible format; we cite Hammes often.

Since all doctrinal publications have many authors and liberally refer—typically without citation—to existing texts, we viewed them as useful summaries or snapshots of the official military take on conventional wisdom rather than as individual positions. U.S. COIN theory is neatly summarized in the Army–Marine Corps counterinsurgency manual, field manual (FM) 3-24. The U.S. Marine Corps *Small Wars Manual* of 1940 reflects both the authors' experiences and U.S. Army material and nearly three decades of articles published on the subject in the *Marine Corps Gazette*. Our third and final doctrinal representation is the 2001 COIN operations manual published by the UK Ministry of Defence.[10] Of the three, we found the British manual to be the most insightful, as well as the most doctrinal. Both the 2006 U.S. COIN manual and the *Small Wars Manual* offer more prescription than doctrine.

Noted COIN scholar Bard E. O'Neill wrote perhaps the most accessible of the various academic works on the subject. In his *Insurgency and Terrorism: Inside Modern Revolutionary Warfare* (1990), we found inspiration for our approach and analysis. O'Neill provided us with a range of pithy quotes, and we liberally reference *Insurgency and Terrorism*. We routinely referred to several other academic works on COIN: Ian F. W. Beckett's *Modern Insurgencies and Counter-Insurgencies* (2001), Stathis N. Kalyvas' *The Logic of Violence in Civil War* (2006), Jeremy M. Weinstein's *Inside Rebellion: The Politics of Insurgent Violence* (2007), Gil Merom's *How Democracies Lose Small Wars* (2003), and Anthony James Joes' *Resisting Rebellion: The History and Politics of Counterinsurgency* (2004). We supplemented this list with articles by

---

Wars was published in 1899. Beckett concurs, stating as much in *Modern Insurgencies and Counter-Insurgencies* (2001, p. 36). *Small Wars* is a useful historical reference.

[10] The manual is actually part 10 of the *Army Field Manual* (UK Ministry of Defence, 2001). A newer version of the manual was reportedly in the works in 2007 but was undergoing further revision as of mid-2009.

Andrew Krepinevich and drew on data from James D. Fearon and David D. Laitin's "Ethnicity, Insurgency, and Civil War" project.[11] We note Ted Robert Gurr's contribution to the study of insurgency and cite his 1974 work on anocracy. We cite Paul Collier's excellent work on civil war and representative government (2000, 2009). Gordon McCormick's quantitative study of insurgency cases helped shape our research design.

How This Monograph Is Organized

This report has six chapters. Chapter Two provides our taxonomy of outcomes that we use to categorize the data. Chapters Three through Six describe our assessments: time and external factors in Chapter Three, internal factors in Chapter Four, and other factors in Chapter Five. Chapter Six presents our findings and elements for policymakers to consider. There are seven appendixes: Appendix A discusses the case studies and our methodology and lists the insurgencies examined in this study; Appendix B provides some supplemental findings; Appendix C describes the multivariate regression analysis; Appendix D lists insurgencies not analyzed for this publication; Appendix E lists the categories used for the spring 2006 survey; Appendix F explains some unavoidable ambiguities; and Appendix G provides the questions used for the autumn 2006 survey.

---

[11] The Fearon-Laitin data also helped us shape our data set. We discuss this process in detail in the appendixes.

CHAPTER TWO

Classifying Outcomes and Selecting Cases

This chapter briefly explains the reasoning that led us to identify four possible outcomes for insurgencies and to identify visible trends particular to each outcome. Understanding these selected outcomes and identifying the inevitable qualitative strengths and weaknesses of the defined terms will arm the reader to decipher the results of this study.

Outcomes

At first glance, classifying outcomes would seem easy: One side—government or insurgent—wins, and the other loses. In practice, outcomes are often more difficult to characterize. For example, if the government gives insurgents amnesty and then allows the insurgent's proxy political party to enter legitimate politics, who has won? Outcomes like this one are frequent in insurgency, so characterizing an individual outcome can be open to dispute.

To conduct this research, it was necessary to define the generally imprecise concept of victory. Answering a subjective question generally requires a subjective answer. Quantitative analysis alone cannot untangle the web of cause and effect intrinsic to insurgency. This means that, for the purposes of this study, we determined what constituted victory. Each case was studied independently, and each researcher determined the outcome for the respective case.

With this caveat in place, it is possible at least to develop general coding classes for outcomes. In this study, we chose four broad classes: government loss, government victory, mixed, and inconclusive or ongoing.[1] While analysts can disagree about how specific individual conflicts should be coded, these four mutually exclusive categories provide reasonable coverage of all possible outcomes in the cases we have examined.

The selection criteria for the 89 cases required that they fall into one of these four categories. Therefore, other insurgencies outside the scope of this study could reasonably be defined by more-specialized terms. More-detailed explanation and justifications for the selection of outcomes are included in the appendixes. The four categorizations are discussed next.

Type I: Government Loss

Most insurgencies fail, since states, no matter how weak or feckless, are typically stronger, better organized, and more professional than nonstate forces.[2] According to one estimate, all of the Latin American insurrections that followed the Cuban model failed utterly—so totally, in fact, that not even their names survived (J. Bell, 1994, p. 115). Successful insurgencies tend to have world historical consequences: Think, for example, of Fidel Castro's victory in Cuba, Mao Tse-tung's in China, the Khmer Rouge in Cambodia, and the mujahideen in Afghanistan.

Insurgencies can win in a variety of ways, including overthrow of the government, successful annexation of independent territory, a marked recognition of minority rights or property rights, or, for the purposes of this study, dramatic political success. Here, we separate a victory from muddled or mixed outcomes by identifying only those insurgencies that effected a political, and thereby a social, upheaval through an existing process. If the current government survived but made some concessions to insurgents, we relegated the case to "mixed outcome."

---

[1]   In a mixed ending, it is relatively easy to see that the campaign ended and that both sides benefited in some way. If we could not identify a clear ending—e.g., some fighting continued but it was unclear whether the violence was tied to the insurgency—then we labeled the case inconclusive.

[2]   According to McCormick, Horton, and Harrison (2006, p. 5), defeat is the condition in which one belligerent is no longer able to "mobilize, transform and employ a diverse array of human and material resources against the other for a strategic purpose."

In the case of insurgent victory, the end game for the state is a rather swift one (see Figure 2.1). Governments, according to McCormick, Horton, and Harrison (2006, p. 6),

> pass a tipping point and begin to decay at an accelerating rate. This is often an indicator that the final period of the struggle has begun. Between the time the conflict enters this phase and the time the state disintegrates, the conflict "speeds up."[3]

What does this mean in concrete terms? As insurgent victory appears ever more probable, a "negative bandwagon" effect takes hold, and previously neutral elements among the population, as well as government supporters fearful of being caught on the wrong side, support the opposition at an accelerating rate. The switching of sides by high-

Figure 2.1
Arc of State Defeat

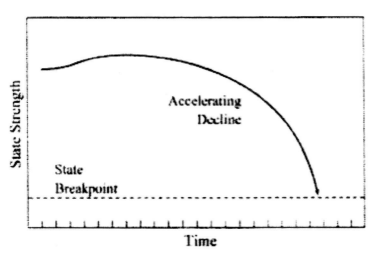

SOURCE: McCormick, Horton, and Harrison (2006).
RAND MG965-2.1

---

[3] The authors' analysis focuses on "insurgent-state dyads"—that is, a conflict between a state and a single insurgent group. As they acknowledge, larger conflicts can include more than one dyad.

level military commanders can be particularly lethal to an embattled regime, as illustrated in Afghanistan in March 1992 when General Abdul Rashid Dostum (and his entire Uzbek militia) defected, thereby dealing a fatal blow to the regime of Mohammad Najibullah.

When insurgents prevail, the end game tends to be foreshortened, and, in many instances, contemporaneous observers did not detect the government's impending collapse until it was already well under way (Beckett, 2001, p. 7). In other words, they failed to identify a tipping point.

In addition to the aforementioned key indicators, other analysts have identified a number of other warning indicators that suggest that a tipping point has been reached and that an embattled regime is in terminal decline. It is very rare that only one or two events will indicate a significant shift in momentum. Typically, intelligence professionals will create a checklist identifying an indication threshold: If a certain number or sequence of events transpires, then a tipping point has likely been reached. These checklists may include a very general set of indicators like the one formulated by Central Intelligence Agency (CIA) analysts (CIA, 1986, p. 11):

- progressive withdrawal of domestic support for the government
- progressive withdrawal of international support for the government
- progressive loss of government control over population and territory
- progressive loss of government coercive power
- capital flight and increasing rates of "brain drain"
- "parking" of financial assets and families of government personnel in safe havens abroad
- increased military desertion rates, particularly among senior officers
- increasing rate of "no-shows" among civil servants, business leaders, and civic leaders
- "drying up" of "actionable intelligence" and other useful information previously supplied by the civilian population.

Type II: Government Victory

In theory, the government wins by destroying the insurgent cadre, the insurgent political structure, or both. However, in practice, governments can and have crushed insurgent forces or movements only to see them reappear years or decades later. This is typically the case when the government fails to address the root causes of the insurgency, as we will show. The government can also achieve victory through legitimate political channels, although this method typically requires at least some concessions to insurgent demands. This is where we drew a fine line between government victories and mixed outcomes.

The end stages of an insurgent loss (or government victory) follow a different trajectory from an incumbent government defeat. According to McCormick, insurgencies typically "decline historically at a *decelerating* rate," as illustrated in Figure 2.2.

While their rate of decline may initially be fairly steep, between the time the conflict enters its end game and the insurgency begins to collapse, the conflict often "slows down" (McCormick, Horton, and

Figure 2.2
Arc of Insurgent Defeat

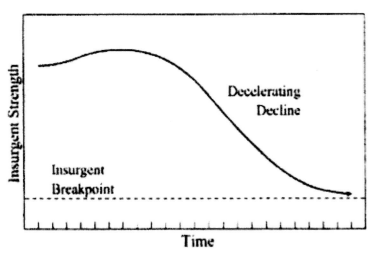

SOURCE: McCormick, Horton, and Harrison (2006).
RAND *MG965-2.2*

Harrison, 2006, p. 6). Government victories are typically not signaled by a dramatic or sudden collapse of the armed opposition: "The insurgency that comes in like a lion . . . may go out like a lamb." Indeed by the time the end comes, "many observers will have assumed it has already come and gone" (McCormick, Horton, and Harrison, 2006, p. 6).

What indicators suggest that a given insurgency is in terminal decline? The CIA (1986) analytical handbook has no list of late-stage markers of successful COIN. However, both the collective body of COIN literature and the results of this study suggest a number of fairly obvious signs that an armed rebellion is likely to decline, or has already begun to do so. First, our key indicators (Byman et al., 2001):

- an increased number of insurgent defections and desertions, particularly among the higher-ranking cadre
- higher volumes of "actionable" intelligence supplied by the population
- the elimination of internal and cross-border sanctuaries and insurgent safe havens.

Additionally,

- "market metrics" suggesting that insurgents must pay more for materiel, services, and information (Krepinevich, 2005a, p. 12)
- a significant drop in international assistance, including financial support from diasporas.

Type III: Mixed (Stalemate/Negotiated Settlement)

Negotiated outcomes are relatively rare. Taking into account a data set broader than the 89 selected cases herein, "[o]nly a quarter to a third of modern civil wars (including anti-colonial wars) have found their way to negotiation" (Zartman, 1995, p. 3).[4] Stalemate—that is, when both sides "are locked in a situation from which they cannot

---

[4]　Naturally, estimates vary: McCormick judges that 20 percent of internal wars result in what he terms "substantive" negotiations.

escalate the conflict with their available means and at an acceptable cost"—provides a critical opportunity for negotiated settlements (Zartman, 1995, p. 8).[5] But such deadlocks seldom occur. Nor do internal conflicts *typically* end because both sides are physically, materially, or politically exhausted. This study shows that exhaustion seems to play only a foreshortening role in the path to negotiation, victory, defeat, or hibernation.

Historical analysis and this study's findings have identified structural and other factors that make internal belligerents unwilling or unable to end the conflict at the bargaining table, including the following:

- long histories of using violence to address political grievances
- the perceived zero-sum nature of internal wars (King, 1997, p. 36)
- unwillingness to forgo opportunities for lucrative plunder (Keen, 1997, pp. 11–12).[6]

The disinclination, even refusal, to negotiate is illustrated in the case of Sri Lanka's Liberation Tigers of Tamil Eelam (LTTE). Here, the cultlike nature of the organization and its leader play a critical role. Most observers concluded that the LTTE's supreme and unquestioned leader, Velupillai Prabhakaran, would never become a "normal" political figure willing to negotiate substantively with his adversaries.[7]

Of course, negotiated settlements have ended other conflicts, such as those in El Salvador, Guatemala, South Africa, and Lebanon. In Northern Ireland, the Provisional Irish Republican Army (PIRA) and the British state reached a mutually recognized stalemate after 25 years of conflict. Intelligence—particularly the extensive use of informants

---

[5] According to Zartman, it is the dramatically asymmetrical nature of internal wars that prevents the development of a stalemate.

[6] As Keen observes, "Conflict can create war economies, often in regions controlled by rebels or warlords and linked to international trading networks; members of armed gangs can benefit from looting; and regimes can use violence to deflect opposition, reward supporters or maintain their access to resources." Such incentives and preferences can obviously encourage the continuation of conflict.

[7] Indeed, he committed suicide when cornered.

within the PIRA—played a key role in neutralizing the organization and creating a mutually perceived deadlock that enabled the peace process.

Type IV: Inconclusive or Ongoing Outcome

This category addresses both those conflicts that have ended with an indeterminate victor and several ongoing insurgencies that are nonetheless deemed statistically relevant for this study. Although it may seem counterintuitive to include an ongoing conflict in a historical analysis, we felt that some ongoing insurgencies both provided excellent data and helped demonstrate the imprecision involved in identifying clear endings.

Determining when an insurgency is over is not as straightforward a task as it might at first appear. Peace is rarely a permanent condition, as exhibited by the waves of protracted violence that followed the apparent "conclusion" of internal conflicts in countries as diverse as Lebanon, Angola, and Afghanistan (and indeed, the United States, which endured years of low-intensity violence after the formal Confederate surrender in 1865). Failure to address the root causes of insurgencies allowed them to hibernate, sometimes undetected, for years before reemerging. Peru roundly defeated the Shining Path movement in 1992 only to see it rise from the ashes ten years later; Peru failed to address the root causes of this mostly rural insurgency.

While identifying precise endings is difficult, it is also hard to reach definitive conclusions about which side won and which side lost. The French ostensibly defeated the Algerian Front de Libération Nationale (FLN, or National Liberation Front) in the 1950s, but, by 1962, Algeria had achieved independence. As discussed earlier, Vietnam is sometimes cited as a COIN success; some analysts point out that, by 1972, the Vietcong had collapsed, and the countryside had been essentially "pacified." However, by 1975, the Vietnamese communists had achieved an absolute victory. In colonial Kenya, the Mau Mau rebellion had been effectively suppressed well before the end of the formal state of emergency in 1959. By 1963, the British had withdrawn and granted Kenya independence from colonial rule.

The Tipping Point: Explanation by Way of Example

We introduced the concept of the tipping point in the first chapter. It figures importantly in our analysis, so we offer an expanded discussion of it here by explaining the concept of the tipping point within the context of a selected case study. The Cuban revolution of the 1950s provides one of the more obvious and accessible vignettes. Cuba not only offers a relatively clear illustration of Gladwell's theory but also hints at the complexities involved in dissecting even well-documented insurgencies.

From the onset of the insurgency against the government of President Fulgencio Batista in 1953, Fidel Castro worked to absorb competing insurgent groups and build grassroots support across the island. However, he was not able to generate a tipping point until the summer of the final year of the insurgency, 1958 (see Figure 2.3).[8]

Figure 2.3
Concept of the Tipping Point

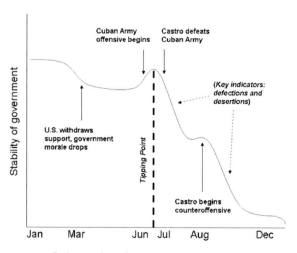

Collapse of the Batista Government in 1958

RAND *MG965-2.3*

---

[8]   Although the Batista regime did not officially abdicate until January 1, 1959, the insurgents had effectively achieved victory by the end of 1958.

President Batista realized that his government was faltering early in 1958. He launched Operation Verano in late June, a last-ditch effort to turn the tide of the war. The Cuban army, consisting of 30,000–40,000 soldiers, sought combat with an insurgent force of perhaps 300 full-time fighters. Despite these odds, by early July, Castro had dealt several well-publicized blows to demoralized army columns while managing to keep his forces intact (Bethell, 1990, p. 452).[9] As it appeared that Castro was gaining the upper hand, insurgent groups not already under his control signed the Pact of Caracas.[10] Castro's success in the field allowed him to consolidate the Cuban insurgency under one umbrella, the penultimate step necessary to forming a new revolutionary government.[11]

The defeat of the army led directly to mass desertions and defections (Bethell, 1990, p. 453). At the end of the summer of 1958, Castro executed a counteroffensive against the Cuban army that led to another round of desertions and defections. By the end of the year, Batista was through (Bethell, 1990, p. 453). The collapse of the army, as clearly indicated by the desertions and defections, identifies the tipping point of the Cuban insurgency in mid-1958.

Insurgencies, like other military conflicts, rarely offer analysts a chain of events leading from a singular identifiable cause or small set of identifiable causes to clear effect. For Cuba and for the remaining cases in this study, we took a broad look at tipping-point dynamics rather than entering into an inescapable logic spiral in the effort to determine causality. In the case of Cuba, one factor in Batista's defeat was the decision by an exasperated U.S. government to withdraw political and military support in March 1958, just a few months before Operation Verano. This left the already demoralized Cuban army further weakened and ripe for collapse (Bethell, 1990, p. 453). The U.S. aban-

---

[9]   McCormick, Horton, and Harrison (2006, pp. 341–346) analyze the Cuban insurgency as a case study and place the tipping point of the campaign in mid-summer 1958.

[10]   The signatories of the Pact of Caracas agreed to align under Castro.

[11]   Bethell and others point out that the military success was tenuous and that Castro was forced to withdraw after striking several major blows against the army. These attacks were sufficient to break the will of the army, even if they did not physically decimate its ranks.

donment of the Batista government was one of many correlating factors leading to Batista's eventual defeat. Castro's military and political actions prior to the 1958 campaigns were also instrumental in effecting Batista's downfall.

Lessons from the Cuban insurgency are drawn both from the tipping point itself and from the clearly identified tipping-point indicators. Even if contemporary analysts could not see the end of Batista's regime in July 1958, the subsequent waves of mass desertions and defections should have provided sufficient warning that the government was on its last legs. In particular, the first cases of officer desertions began to manifest after the army was defeated in the summer. By providing a rather clear linkage between a set of actions and indicators, the Cuban case also provides salient lessons for intelligence experts.

Key Indicators: A Note of Caution

Desertion and defection rates are useful indicators, but, as with any analytic or predictive tool, they are not foolproof and cannot be used as independent metrics. Insurgencies develop and end according to a network of actions, reactions, inactions, and happenstance, all of which in turn create interconnected shifts in ground truth. In the case of Cuba, desertions and defections proved useful in identifying the tipping point with relatively little ambiguity. The case of the Chieu Hoi program, on the other hand, provides a cautionary tale.

"Chieu Hoi," or "Open Arms," was implemented as a combined program between the United States and the government of Vietnam (GVN) in 1963 lasting through 1971. It was designed both to encourage defections and to help reeducate and integrate defectors. The program appeared to be such a success that the GVN established a separate Chieu Hoi ministry in 1967 (Koch, 1973). In the year following the 1968 Tet Offensive, the program reported that the number of defectors increased year-on-year from approximately 18,000 to more than 40,000 (Koch, 1973, p. 11). Many defectors cited the Vietcong's inability to hold terrain in the wake of Tet as a primary motivator to switch sides. Based on casualty counts, intelligence reports, and, in part, defection rates, many contemporaneous observers viewed Tet as an operational success and believed that it offered a window of opportunity for the United States and the GVN. Taken in isolation, it appears, even in retrospect, that the sharp increase in defections reflected an operational shift in momentum from the Vietcong to the combined U.S. and GVN forces.

However, simultaneously with the decay of Vietcong influence and the surge in defections from mid-1968 to 1969, U.S. political willpower crumbled. U.S. troop levels began to ebb in late 1968 as grassroots opposition to the war swelled. President Lyndon Johnson subsequently declined to run for another term in office,[12] and the communist north overthrew the GVN in 1975. In our brief case study of the Vietnam War, we argue that the Tet Offensive marked a tipping point signaling the beginning of the end for the GVN.

In the case of Vietnam, the obvious tactical indicators were therefore misleading: They obscured a shift toward strategic defeat.[13] The fallacy of reliance on a single metric or narrow set of intelligence indicators is revealed in other cases as well. The key indicators derived from the quantitative analysis in *How Insurgencies End* are intended to inform rather than direct the predictive analysis process in a COIN campaign.

---

[12] The official White House biography on President Johnson (White House, undated) cites the crisis in Vietnam as a central factor in his decision to withdraw as a candidate.

[13] Ian F. W. Beckett (2001, p. 198) points out that many of the Chieu Hoi defectors were simply looking for money or a temporary break from fighting, another justification for caution when interpreting indicators.

CHAPTER THREE

Assessments of Insurgency Endings: Time and External Factors

Assessments explaining the key findings and indicators are presented in this and the following chapter. Each assessment is supported by one or more examples drawn from the broader selection of 89 cases and previous case studies researched at RAND (Rabasa, Warner, et al., 2007). In this way, the quantitative data are anchored in qualitative detail. The data resulting from the survey process are derived from the knowledge and opinions of the individual researchers participating in the survey.[1] Therefore, the assessments of insurgency endings are inexact. Even though functional-area experts carried out the study, the somewhat subjective nature of this research exposes *How Insurgencies End* to legitimate criticism.

It is debatable that a phenomenon as complex and diverse as insurgency could be explained through purely scientific process; subjective assessment must play a role to a great degree. Instead, the assessments are offered with the intent both of providing a useful interpretation of the history of insurgency and of sparking debate over conventional wisdom.

The quantitative portion of this study reinforces the notion that insurgency-related metrics are inherently imprecise. As obvious as this may seem to some, senior staffs in Iraq and Afghanistan continue to rely on quantitative metrics to gauge relative success, alter plans, and allocate resources. This in and of itself is a phenomenon worthy of con-

---

[1] Anthony Butera, Peter Chalk, Sara Daly, Brian Jackson, Seth Jones, Martin Libicki, Bill Rosenau, Paraag Shukla, and Anna-Marie Vilamovska formed the research staff that participated in the surveys.

tinued exploration, especially considering the challenges presented in conducting a quantitative analysis of readily accessible and relatively static historical data on insurgency and COIN.

These epistemological challenges are arguably unique to the study of warfare. There are intangible factors common to all forms of conflict, including those of political will and popular support. However, state-on-state conflicts usually provide historians and analysts with an ample set of well-codified data. These data include types and amounts of equipment produced, numbers of soldiers fielded, production capacity, and generally accurate casualty figures. Insurgencies rarely generate information to this level of detail, and, when they do, the data are often irrelevant or misleading. Casualty levels that would normally decimate or break a standing army often prove insufficient to drain the recruiting pool or crush the will of an insurgent cadre. Furthermore, the effort and sacrifices necessary to create these casualties might prove at least partly causal in an eventual government defeat.

In judging the results of *How Insurgencies End*, it is also worth observing that *correlation is not the same as causation*. Outside support to insurgents, for instance, correlates with insurgent victory. This does not necessarily mean that such support spelled the difference between victory and defeat (i.e., actually *caused* it), nor does outside support necessarily cause an insurgency to succeed that was not on the road to victory anyway.

Some, but not all, of the data in this chapter provide relatively clear findings. As a rough rule of thumb, for the sample sizes in question, it takes roughly a 20-percent difference in outcomes before having confidence that something more than random variation is at work.

Assessments of Time and External Factors

Duration of Conflict

Conventional wisdom states that insurgencies last about ten years (see Figure 3.1).[2] In this, our first finding, conventional wisdom essentially bore out. If they survive the "proto" stage, insurgencies generally last several years from beginning to end; the median length is ten years, but with long tails (see Table 3.1).[3] There is a good chance that insurgencies can be concluded within 16 years, but, if the insurgency survives

Figure 3.1
Insurgency Durations and Outcomes

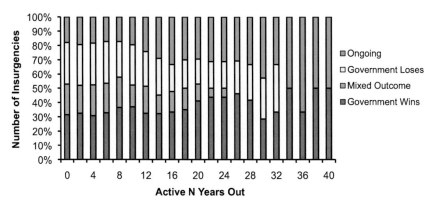

NOTE: The number of insurgencies in each category was normalized to 28 (the number of insurgencies in the government-won category) by arraying insurgency lengths in monotonic order and then inserting interpolated results evenly throughout the range. This method does not significantly change the percentile rankings of the results.
RAND *MG965-3.1*

---

[2] That a ten-year timeline reflects conventional wisdom is a subjective assessment of the researchers. In discussions on the Iraq and Afghanistan wars, a range of COIN experts, including General Petraeus, have referred to a timeline that ranges from five to 15 years. In a 2007 interview with Fox News' Chris Wallace, General Petraeus stated that insurgencies last "nine or ten years."

[3] A "tail" period takes place typically after major combat operations have ended and the insurgency is in its death throes or in seclusion. Tail periods are often the most tenuous for counterinsurgents as they try to balance reforms, troop levels, negotiations, amnesty programs, and reintegration of former insurgents.

Table 3.1
Surviving Insurgencies After Up to 40 Years

Outcome	Years Out																				
	0	2	4	6	8	10	12	14	16	18	20	22	24	26	28	30	32	34	36	38	40
Government wins	28	25	20	19	19	17	12	10	7	7	7	7	7	6	5	2	2	2	1	1	1
Mixed outcome	19	15	14	12	11	7	7	4	3	3	2	1	1	0	0	0	0	0	0	0	0
Government loses	26	22	19	17	13	13	9	8	4	4	3	3	3	3	3	2	2	0	0	0	0
Ongoing	16	15	12	10	9	9	9	9	7	6	5	5	5	4	4	3	2	2	2	1	1

that long, likelihood of an expeditious conclusion from then on tends to decline.

Once an insurgency starts its third decade, the government takes longer to win it than to lose it. However, these data are not conclusive, since they refer to several ongoing conflicts. They do show that the average length of insurgent-won conflicts does not exceed the average length of government-won conflicts. Therefore, it is safer to conclude that *insurgents do not necessarily win as long as they manage to "hold out."* So, while this finding reinforces the notion that insurgencies last about ten years, it also flies in the face of the insurgent "long war" theory of Mao Tse-tung and Ho Chi Minh. Mao (1961 [2000], p. 27) believed that, once an insurgency gained the support of between 15 and 25 percent of the population (a "decisive figure"), it essentially became invincible:

> Historical experience suggests that there is very little hope of destroying a revolutionary guerrilla movement after it has survived the first phase and has acquired the sympathetic support of a significant segment of the population.

Guillén (1973, p. 232) subscribes to Mao's long-war theory but believes that longevity simply favors the steadfast, either insurgent or counterinsurgent:

> In a war of liberation the final victory is not decided by arms, as in imperialist wars. In a revolutionary war that side wins which endures longest: morally, politically, and economically.

Perhaps less confident than Mao in the strength of the insurgent vanguard, Guillén (1973, p. 241) believed that the insurgency must have the support of the "great majority" of the people in order to win.[4] Hammes (2006, p. 183) states very plainly that longevity favors the

---

[4]   On p. 253, Guillén states that the insurgents can win even when facing a 1:1,000 ratio of insurgents to counterinsurgents, as long as the insurgents have the support of 80 percent of the population.

insurgent. Here, he describes one of the strengths of the Iraqi insurgents circa 2004:

> The greatest strength of the insurgent is the fact that he doesn't have to win. He simply has to stay in the fight until (the coalition) gives up and goes home. By simply not losing, [insurgencies] compel their opponent to choose—either continue to fight, perhaps indefinitely, or quit and go home.

Galula (1964 [2006], p. 12) makes the same case rather succinctly: "Because of the disparity in cost and effort, the insurgent can thus accept a protracted war; the counterinsurgent should not." Kilcullen (2009, pp. 95–96) demurs, suggesting that the counterinsurgent must be the one to hold out for the long haul. Here, he quotes a U.S. Army colonel who discusses a long-term road-building project his unit undertook in Afghanistan:

> [W]hen you mix this sense of long-term commitment with a persistent-presence methodology, it becomes apparent to everyone that [we] are going to be in the towns for a long, long time. The U.S. isn't going away tonight and leaving the elders to cope with the Taliban on their own.

However, Kilcullen and others have rather vocally doubted U.S. strategic patience. We found the lack of patience to be an endemic problem among external sponsors of COIN operations.[5] How, then, can it be true that governments (which, like insurgencies, typically rely on some form of external sponsorship) may be winning more often in the long run? Considering Mao's caveat, many of the insurgencies defeated in the long run (after about 20 years) failed because they were not able to sustain popular respect and support. We found that they often lost this support because (1) they went "off message" by redefining their goals in a way that alienated the populace; (2) they used indis-

---

[5] Ulterior motives of some external sponsors undermined many government COIN campaigns.

criminate terror tactics; or (3) the government undermined the theme of the insurgency with offers of social and political reform.[6]

Sri Lanka and the LTTE. It is also interesting to look at cases in which the insurgency retained a sizable public following but still suffered military defeat. The conflict between the government of Sri Lanka and the LTTE provides an interesting perspective on this dynamic and on the ebb and flow of those insurgencies that last beyond the ten- to 16-year "hump" (see Figure 3.2). For the duration of the conflict, they managed to sustain the commitment of at least a sizable portion of the 15-percent-Tamil minority in Sri Lanka as well that of a very sizable and vocal expatriate community.[7]

Figure 3.2
History of the Tamil Tiger Insurgency

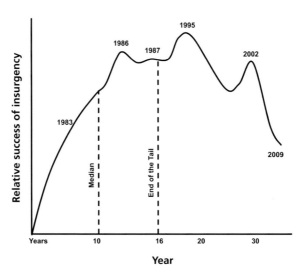

RAND MG965-3.2

---

[6] We will show in our findings that the latter case is also infrequent and unlikely to succeed as a sole causative factor in defeating an insurgency—it is often more difficult to effect reform once an insurgency is under way.

[7] Various reporting indicates that the LTTE press-ganged (forced) children into its ranks and expropriated funds from citizens. It is difficult to gauge the group's actual popularity over the course of the conflict, especially taking into account popularity within diasporas and Indian Tamil communities.

The LTTE was formed in the 1970s and began fighting an aggressive and efficient insurgency campaign circa 1983.[8] As the insurgency approached the ten-year median (1987), two attempts were made to end the conflict. However, peace talks held in 1985 and again in 1987 failed to satisfy the LTTE or Tamil politicians, and the insurgency continued.[9] In 1994, peace talks were rekindled. This initiative occurred two years shy of the 16-year mark. When these talks also failed, the insurgency slipped into a period of entrenchment marked by intensive flare-ups of violence and the introduction of a systematic suicide-bombing campaign by the LTTE. At this point, the LTTE had made it "over the hump," with the chances for swift resolution to the conflict seeming to dwindle after 16 years. Indeed, little to no progress toward resolution was made during the eight years subsequent to the failed 1994 talks.

In 2002, 26 years into the conflict, the government of Sri Lanka conceded to many of the LTTE demands. The Tigers were granted rule over a semiautonomous zone in the Jaffna peninsula. It appeared that the LTTE, with a sizable military force and semiautonomous rule, had reached a tipping point toward what it might perceive to be a victorious end state. However, it continued to press its attacks. Perhaps stiffened by LTTE recalcitrance, the Sri Lankan military recovered, reorganized, and had begun a concerted attack against the LTTE across the Jaffna peninsula by 2007; the government had at least temporarily reversed the course of the conflict. The LTTE was driven back, suffering significant losses. As this study concluded, the military wing of the LTTE had been decimated, and Velupillai Prabhakaran, the group's leader, had been killed or had committed suicide.

In its 34th year, the LTTE insurgency appears finished, but the *conflict* likely has not ended. It is possible that the government has generated a favorable tipping point with its latest military campaign.

---

[8]   The timeline for this general summation of the LTTE insurgency was taken from various sources, including Ethirajan (2009).

[9]   Some reports indicate that the LTTE hierarchy killed moderates within their own movement to deliberately undermine these talks. Brendan O'Duffy viewed LTTE engagement in the peace process as a tactic solely designed to provide breathing space for the development of guerrilla and conventional capacity (O'Duffy, 2007, p. 262).

Statistically, however, this far into the conflict, the government has only a slightly greater chance of completely defeating the insurgency—in whatever form it takes—than succumbing to eventual defeat or an inconclusive negotiated settlement (see Figure 3.3). As of mid-2009, it was unclear whether the government saw the necessity to address the underlying concerns that first instigated the insurgency.

Qualitative analysis indicates that the most likely end-state sequence at this stage will involve protracted negotiations, amnesties, political engagement, and the development of at least one extremist splinter group or group of hibernating insurgent cells. This sequence of end-state events may play out over the course of an additional five to ten years.

The two sides may negotiate for several years before a comprehensive peace treaty is signed. With the death of Velupillai Prabhakaran and the capture of his second-in-command, the Sri Lankan government might fail to identify a valid interlocutor, and it may waste time engaging with a de jure rather than de facto representative. The govern-

Figure 3.3
Hypothetical Trajectory of Tamil Tiger Insurgency: Notional End-State Sequence, Assuming Government Victory

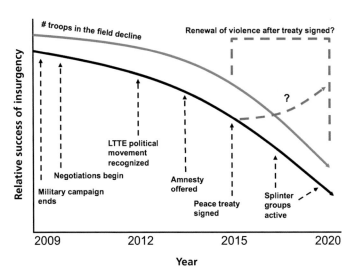

ment will probably be required to keep a sizable number of troops on the Jaffna peninsula to preserve stability, although financial and political pressure may force the government to conduct a phased withdrawal as negotiations progress.[10]

As shown in the cases of Northern Ireland and Algeria (1992–present), the juxtaposition between successful negotiations and a decreased security-force footprint presents the greatest opportunity for splinter groups to renew insurgent activity. In their weakened state, these splinter groups are most likely to resort to terrorist attacks, particularly in the case of an organization like the LTTE that is accustomed to using suicide bombings to good effect. If the government proves incapable of legitimately addressing Tamil humanitarian and political concerns during the course of the negotiations (behaving as an anocracy or autocracy) and fails to maintain adequate security, these splinter groups may gain a foothold and reignite a full-blown insurgency.

We noted one qualitative marker of success or failure associated with these splinter groups. If this irredeemable core retained the support of the populace, either it conducted very low-level operations until it rebuilt capacity, or it fell off the map ("hibernated") for a time but eventually reemerged and reengaged in violence. If it did not retain the support of the populace, it became, in essence, a small terrorist group whose elimination could ultimately end the insurgency. This latter case was possible only when the government had addressed the root causes of the conflict.

Sanctuary Available

Although a few 1960s-era insurgency theorists underplayed the necessity for sanctuary, Mao, Giáp, and most modern COIN theorists concur

---

[10] This notional scenario is provided to show one version of an "arc to ending" for a government victory. It was derived from both the quantitative and qualitative portions of this study. There are insufficient accurate data to determine the average length of a negotiation in the case of a government victory, due to the possible range of involved actors and phases of negotiations. As of mid-2009, it appeared that the Sinhalese government had taken a very aggressive tack and was actually increasing the number of troops it was putting into the field. It was also exacerbating ethnic and religious tensions with the Tamils instead of attempting to address their social concerns.

that insurgent sanctuary correlates with insurgent victory.[11] Our finding on sanctuary is consistent with conventional wisdom: The large sample size and wide divergence in statistical outcomes indicate a close correlation between sanctuary and ending (see Figure 3.4). Further, the data were clearly reinforced by the qualitative research conducted not only by our analysts but also by many of the theorists and practitioners cited in this study. In our study, insurgents who have enjoyed sanctuary have won almost half of the conflicts that have been clearly decided (23 out of 52). Only three of the insurgencies that operated without sanctuary ended favorably for the insurgents (three out of 22, with others ongoing or mixed). An additional finding from the analysis and survey is noteworthy: If sanctuary was involuntary (that is, the receiving state had no great desire to shelter insurgents but had little means to stop

Figure 3.4
Value of Sanctuary

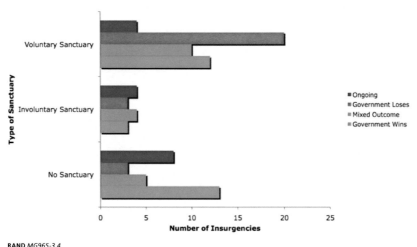

RAND MG965-3.4

---

[11] McCuen (1966) states that "sanctuary" in the form of secure base areas is also critical to the counterinsurgent. We did not explore his hypothesis in this study. McCuen introduces the concept on page 54 of *The Art of Counter-Revolutionary War* and dissects it in detail through the course of the book.

them), the insurgents did not do better than average.[12] This suggests that sanctuary may be more an indicator of broader and hence more-valuable state support than of a safe locale.

The total absence of sanctuary leaves insurgents with only a one-in-seven chance of winning (out of decided cases). While this finding seems to be reinforced by conventional wisdom, not all of our noted authors agree on the subject of sanctuary. Here, Mao Tse-tung (1961 [2000], p. 107) identifies the basic conditions necessary to establish a "base," and describes the value of such a base to the insurgent:

> A guerrilla base may be defined as an area, strategically located, in which the guerrillas can carry out their duties of training, self-preservation and development. Ability to fight a war without a rear area is a fundamental characteristic of action, but this does not mean that guerrillas can exist and function over a long period of time without the development of base areas.

We argue that Che Guevara provides an outlying opinion. He recognizes the need for some "inaccessible positions" to preserve the insurgents in the early stages of the conflict, stating, "It is essential always to preserve a strong base of operations and to continue strengthening it during the course of the war" (1969 [2008], pp. 17, 19, 78).

But Che (1969 [2008], p. 119) later implies that sanctuary and static bases lead to stagnation. He warns against relying on foreign sponsors and romanticizes the peripatetic nature of Mao's insurgency, perhaps misconstruing the narrative of the Long March.[13] Che (p. 49) describes zones of security and zones of warfare and goes into great detail describing "unfavorable ground" for insurgencies (plains and

---

[12] Sri Lanka (LTTE) and a few other cases demonstrate the value of noncontiguous sanctuary: Coethnic Tamils fostered the growth of the LTTE from safe haven in the Tamil region of India. We did not distinctly code cases of noncontiguous sanctuary.

[13] "Unconditional help should not be expected from a government whether friendly or simply negligent, that allows its territory to be used as a base of operations; one should regard the situation as if he were in a completely hostile camp." Dick Wilson's (1971) *The Long March, 1935: The Epic of Chinese Communism's Survival* is an excellent if narrow resource on the Long March and on Maoist insurgent practice.

suburban areas), but, for his *focos* to thrive, the vanguard of the insurgent army must remain detached from secure, fixed sites:

> [The insurgent's] life is the long hike. . . . [T]he guerrilla band moves during daylight hours, without eating, in order to change its position; when night arrives, camp is set up. . . . The guerrilla fighter eats when he can and everything he can. . . . His house will be the open sky.

Che's theories applied well in the mountains of Cuba but nowhere else.[14] His confederate, Debray (1967, p. 45), adds,

> [T]he advantages a guerrilla force has over the repressive army can be utilized only if it can maintain and preserve its mobility and flexibility. [By] going over to the counterattack . . . it catalyzes the people's energy and transforms the foco into a pole of attraction for the whole country.

Debray's comments give some insight into the inherent vulnerabilities of the *foco* model. Both Che and Debray envision an insurgency that is typically on the move and shuns the recurring periods of isolation that allowed Mao Tse-tung's Chinese communist insurgents to survive their darkest days: The Long March ended at an internal sanctuary.

Urban insurgency dogma also flips Maoism on its head. Marighella (2008, p. 34) eschewed the reliance on sanctuary and fixed bases:

> Urban guerrillas . . . are not an army but small armed groups, intentionally fragmented. They have neither vehicles nor rear areas. Their supply lines are precarious and insufficient, and they

---

[14] See, for example, Hammes (2006, p. 77): "[Che] did not understand that their success was based on the unique conditions of Cuba, which included a pending collapse of the Batista regime. Che paid for his mistaken theory when he tried to apply it (elsewhere)." Che and Debray tend to speak with one voice on this issue.

have no fixed bases except in the rudimentary sense of a weapons factory within a house.[15]

While he accepts the need for some fixed sanctuary in the latter stages of a campaign, urban theorist Abraham Guillén sees little need for bases in the early and middle stages of an insurgency:

> Strategically, a very small guerrilla army must operate in view of bringing about a mass insurrection without engaging the popular forces in an initial battle, without fastening itself to a given space, without creating fixed mountain encampments.[16]

Of course, in nearly every case of urban insurgency, the absence of sanctuary was debilitating; the statistics caught up with Marighella and most of his contemporaries. General Giáp and Ho Chi Minh ascribed to and practiced traditional Maoist insurgent doctrine: "In the course of the national liberation war, the building of bases for a steadfast and long resistance was an important strategic question" (Võ, 1961 [2000], p. 110).

Giáp later goes on to describe in detail the value of strengthening the rear area in supporting the third phase of insurgent warfare. COIN conventional wisdom also recognizes the value of sanctuary to the insurgent, although we found that the subject received surprisingly little attention from our recognized experts. Galula, like Mao, envisions sanctuary as a stepping-stone from which the insurgency progresses from phase to phase. To forestall this eventuality, the counterinsurgent should use economy-of-force techniques to raid "other" areas. The counterinsurgent does so in order to "prevent the insurgent from developing into a higher form of warfare, that is to say, from organizing a regular army. This objective is fulfilled when the insurgent is denied safe bases" (Galula, 1964 [2006], p. 81).[17]

---

[15] He contradicts these thoughts in other writings, but this is his contribution to conventional wisdom.

[16] We noted earlier that Guillén also thinks little of urban safe houses.

[17] His recommended technique for denying sanctuary—the targeted raid—has very narrow applicability and would not prove either feasible or effective in a variety of the cases we stud-

O'Neill (1990, p. 119) gives the subject proper treatment. He first states that sanctuary is a prerequisite if the insurgency is to expand to Maoist third-phase operations:

> Without a contiguous sanctuary, groups such as the Tamils in Sri Lanka . . . are handicapped severely when it comes to expanding their small-scale guerrilla attacks to large, sustained, and widespread guerrilla campaigns.

He also implies a strong correlation between sanctuary and insurgency endings, stating that, in the absence of sanctuary, the insurgents

> must depend on the hope that government ineptitude and demoralization in the army will eventually result in political abdication. If the government and army do not falter, the lack of an adjacent sanctuary can be a glaring, if not fatal, deficiency. (pp. 117–119)

O'Neill (1990, p. 56) addresses the fact that fixed bases may provide the counterinsurgent with the opportunity to exploit a critical vulnerability of the insurgency.

The 2006 version of FM 3-24 (p. 1-16) describes sanctuary, but it fails to address the practical value of sanctuary to the insurgent. Worse, it does not definitively state that denying sanctuary might give the counterinsurgent an operational or strategic advantage. Better discussions on the subject can be found in the myriad case studies written about Vietnam, Afghanistan, Malaysia, Thailand, Mozambique, Guatemala, and Nicaragua.

Statistics and conventional wisdom aside, the value of voluntary sanctuary is fairly self-evident. With a secure space, insurgents can train, organize, rest, refit, and, if necessary, hibernate. We suggested that voluntary sanctuary implies greater benefit for the insurgent, probably in the form of active assistance by the host country. This proved

---

ied. For example, any raid against Syria or Iran of sufficient size to deny sanctuary to AQI would have escalated the Iraq war and undermined U.S. theater objectives. Efforts to deny or interdict sanctuary through the use of indirect or remotely directed fires appear to be a questionable tactic, as evinced by U.S. strikes into the federally administered tribal area (FATA) in Pakistan.

to be true in several, but not all, of the 46 cases of voluntary sanctuary we identified (e.g., Moroccan Polisario in Algeria, and, conversely, the Algerian FLN in Morocco, Tunisia, and Libya). There were many cases in which insurgent presence simply was tolerated—i.e., the host country provided sanctuary willingly (or perhaps grudgingly) but gave no additional resources or support to the insurgents. We classified these cases as voluntary but did not draw a distinction between "voluntary/supporting" and "voluntary/tolerated." Although we did not code for tolerance in our quantitative findings, we find it to be a helpful qualitative distinction during our broader analysis.[18]

Involuntary sanctuary is less beneficial to the insurgent than voluntary sanctuary. The absence of "voluntary" support may mean that the insurgents are restricted to operating in inhospitable terrain, perhaps in remote jungle or mountain areas. Without logistics and lines of communication, isolation in these remote areas can severely stunt the development of an insurgent cadre. Five of the 14 insurgencies with involuntary sanctuary also lacked external support, while three received only limited support. In other words, about half of involuntary-sanctuary cases were also effectively unsupported. Our research showed that a lack of sponsorship correlated with a 3:18 win-loss record (considering only decided cases). Projecting these results onto prospective conflicts, it would seem that involuntary sanctuary is not a critical element in insurgent victories.

Iraq and AQI. Iraq is an interesting case in that it offers an opportunity to analyze the effects of both external and internal sanctuary (see Figure 3.5). In particular, external Syrian support to Iraqi insurgents reveals subtle gradation between the lines of voluntary and involuntary sanctuary. Syria reportedly has harbored former Iraqi Ba'ath party leadership since the onset of hostilities in 2003 and has allowed former regime and nationalist groups to conduct logistics operations across its border and into Iraq. According to media reports, however, Syria has not allowed these insurgents to establish firm bases of operation within its borders, and Syria is not necessarily providing *active* support (i.e., logistics or money) to any great degree. Between 2003

---

[18] O'Neill (1990, p. 118) lists "tolerance" as a kind of sanctuary.

Figure 3.5
Al Qaeda in Iraq Insurgency in Anbar Province

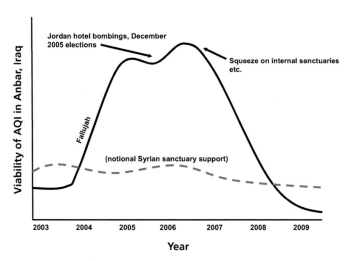

NOTE: AQI = al Qaeda in Iraq.
RAND *MG965-3.5*

and 2009, Syria has variously supported, allowed, overlooked, or (more recently) obstructed the movement of foreign fighters into Iraq. At the same time, elements within the Syrian regime have probably tried to eliminate foreign-fighter safe houses, but the government does not necessarily speak or act consistently or unanimously. Therefore, we can say that Syria is providing (at least) involuntary sanctuary to some foreign fighters.

Between March and November 2004, various nationalist insurgent groups and AQI divided control of the city of Fallujah.[19] The city was a denied area for U.S. forces throughout the entire summer of that year. Ongoing negotiations left the insurgents relatively unmolested in Fallujah until at least the middle of the summer, and AQI took advantage of the physical sanctuary afforded by the city to establish itself as

---

[19] For the sake of simplicity, we have generalized the notion of "nationalist" groups and AQI. At least ten nationalist groups operated in Iraq at various times (e.g., 1920 Revolution Brigade, Jaish Islami, Harakat al-Islamiyah, Ansar al-Sunnah). And, in reality, AQI took on various forms over the course of the conflict and was itself divided into several separate, but typically complementary, entities.

the dominant insurgent organization in Iraq. By the time the United States retook the city in November 2004, the credibility of the coalition had been severely strained. AQI thrived in Anbar province until early 2007, relying not only on "rat lines" into and out of Syria but also on the remaining internal sanctuaries within the jazira, or isolated desert areas, along the shores of Lake Thar Thar, in rural areas southeast of Fallujah, and in the rural terrain surrounding the provincial capital, Ramadi.

In late 2006 and early 2007, grassroots momentum against the group materialized in the form of anti-AQI tribal militias. At approximately the same time, the United States conducted its "surge," increasing its presence in Iraq and giving U.S. forces in Anbar province the flexibility to interdict various internal sanctuaries. Coalition troops effectively sealed off Ramadi and began to make aggressive forays into the rural havens. Growing confidence in Iraqi security forces and tribal militias enabled marines and soldiers to deploy in distributed outposts from which they were better able to control lines of communication and to deny key terrain to the insurgents. Within a year, AQI had been squeezed out of Anbar. Attack levels dropped precipitously. AQI-affiliated insurgents probably maintain some low-level presence in various towns and cities along the Euphrates River circa mid-2009, but the first iteration of the AQI insurgency in Anbar effectively ended between 2007 and 2008.[20]

We correlate the existence of "mixed" sanctuary in Syria with the ability of AQI and other insurgent groups to operate in Anbar province for approximately three to four years. Because so many other factors contribute to any analysis of the Iraqi insurgency, we also judge this correlation to be somewhat ambiguous. While Syrian sanctuary helped sustain the insurgents, it is unlikely that the loss of this sanctuary would have ended or even severely restricted the insurgency, as it did in many other cases (e.g., Greece, Thailand, respectively).

---

[20] This "ending" is, of course, localized and does not necessarily signify the end of AQI or of other insurgency movements in Iraq. In fact, many AQI insurgents who fled Anbar settled in Mosul or in other areas north of Baghdad, where they remained active as of mid-2009.

We believe that the correlation between the loss of internal sanctuary and the defeat of AQI in Anbar is much stronger. Anbari insurgents were availed of a nearly limitless supply of weapons and munitions scattered about the countryside in prewar bunkers and postinvasion caches. Because so many insurgents were also former soldiers, training requirements were few. The "insurgency" was a dissociated series of violent movements practicably incapable (we argue) of transitioning into third-phase warfare and therefore not requiring a secure base of operations. In essence, they needed only physical sanctuary to plan, rest, refit, and build homemade explosive devices—tasks easily accomplished in any garage or rural home. When internal sanctuaries were squeezed, through both coalition action and grassroots activity, the insurgents were forced to disperse, flee, or join the government security services.[21] As Kilcullen (2009, p. 145) puts it,

> [The] point of the operations was to lift the pall of fear from population groups whom terrorists had intimidated and exploited, win them over, and work with them in partnership to clean out the cells that remained—as occurred in Al Anbar Province in 2006–2007 and later elsewhere in Iraq as well. The "terrain" being cleared was human terrain, not physical terrain.

While it is difficult to correlate the availability of sanctuary directly with insurgent success, it is fairly simple to show correlation (not necessarily causation) between the loss of sanctuary and defeat. A few cases are sufficient to illustrate the point that sanctuary—voluntary, involuntary, or internal—is a fundamental provision for the insurgent. Iraq provides an interesting (yet, so far) inconclusive example. Greece (1945–1949) presents a stronger, if not clear-cut, demonstration of our finding. We touch on this case and complete this section with a brief comment on the ongoing conflict in Afghanistan.

---

[21] We can only infer correlation between the loss of internal sanctuary and the defeat of AQI in Anbar. Many other known and unknown factors fed into these end-game dynamics, and a definitive history of this period had yet to be written as of mid-2009. Contemporary literature attributes the events of this period to the development of the Awakening movement or to the "surge." Kilcullen gives credit to the increase in forces but also to shifts in U.S. tactics and a strategic reassessment that led to a new population-centric campaign plan.

Greece and the Greek Communist Party. The Greek Communist Party failed to seize control of Greece in the waning stages of World War II and then was dealt a crushing postwar political defeat. Seeing no legitimate path to power, the communists renewed attempts to overthrow the Greek government in 1945 by means of its military wing, the Democratic Army of Greece (DSE). Communist political operatives and DSE fighters enjoyed voluntary sanctuary in Yugoslavia and, to a lesser extent, Albania, through the middle of 1949. Both countries provided the Greek insurgents with external support in the form of weapons, training, and, in some cases, funding and political assistance, but Yugoslavia was far better positioned to sustain the insurgents. With reliable sanctuary and external support, and in the absence of credible or sufficient government security forces, the insurgency expanded both in size and in scope through mid-1948 (Murray, 1962).

In late 1947, the insurgents decided to shift to conventional light-infantry tactics and started to hold ground in mountain redoubts within the Greek border. Army columns attacked the DSE insurgents throughout 1948—in some cases, driving them out of these internal bases. However, the insurgents would melt away to sanctuary, quickly reconsolidate, and then cross back into Greece to establish new battle positions (see Figure 3.6). The Greek army continued to founder until the beginning of 1949 despite nearly doubling in size and receiving rather substantial assistance from the United States.

By early 1949, however, the army had considerably improved and, by springtime, had taken the insurgents to task on several occasions. By the early summer, the insurgents were on the defensive but still retained significant influence with the population (both positive and coercive) and could rely on extensive militia reserves. Everything fell apart, however, when Marshal Tito closed the borders of Yugoslavia to the insurgency in July 1949.[22] Separated from its primary sanctuary and main source of external support, the remaining insurgent cadre stumbled back into Albania. Although Albania continued to provide

---

[22] Tito, both marshal and president, closed his borders when the Kommounistikó Kómma Elládas (KKE, or Communist Party of Greece) decided to retain ties with the Soviet Union in the wake of the Soviet-Yugoslav split.

Figure 3.6
Trajectory of Greek Insurgency

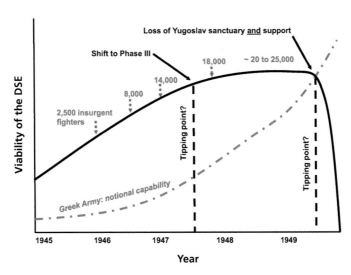

RAND *MG965-3.6*

some support, without the voluntary sanctuary and external sponsorship of Yugoslavia, the movement was doomed. The war ended in late 1949 (Shrader, 1999).[23]

The case of the Greek insurgency provides no exceptions to the rule of complex correlation and causation. That the DSE insurgency ended is not seriously contested, but there is some debate over the most prominent correlative factors.[24] We found that there tends to be some broad agreement over three of these:

- *The insurgents decided to play on the army's ground.* By moving to third-phase warfare, the insurgents offered the Greek army an opportunity to deal a conventional blow to the insurgent infra-

---

[23] Shrader (1999, p. 260) describes the effect that Tito's repudiation had on the DSE, as well as the other myriad factors that can be correlated with the defeat of the DSE and KKE.

[24] Some authors stress that a muddled political agenda and a coincidentally disjointed propaganda campaign buckled the foundation of the insurgency to the point that its collapse was all but inevitable. McCuen (1966, pp. 299–304) provides a detailed analysis of the Greek case.

structure. Organized as light infantry, the insurgents were out-gunned and outmatched in the end (Galula, 1964 [2006], p. 12).[25]

- *The army got its act together.* By early 1949, the army began to pull together and perform in the field. The appointment of a new, aggressive, and well-respected commander did wonders for morale and organization. By mid-1949, the army was more than capable of defeating a light-infantry force one-tenth its size.[26]

- *The Yugoslav sanctuary was indispensable.* Most insurgent logistics flowed through Yugoslavia, and, when they lost the Yugoslav sanctuary, the insurgents were cut off and starved of supplies, cash, ammunition, weapons, and political patronage. Albanian sanctuary and support was utterly insufficient to sustain the insurgency.

No matter how these three correlating factors are weighted in comparison to each other, the denial of Yugoslav territory and support played an essential role in the ultimate collapse of the insurgent cadre. Sanctuary—whether lost or held—is closely correlated with some uncharacteristically dramatic insurgent endings and a number of all-too-characteristic government defeats (see Table 3.2).

Afghanistan, Pakistan, and the Taliban. As with the case of Iraqi sanctuary in Syria, that of the Afghan Taliban sanctuary in Pakistan is a bit more complex than it first appears. Although the Taliban direct their messages and attacks against the Islamic Republic of Afghanistan, the insurgency has always reflected cross-border ethnic and tribal influences. The Taliban comfortably operate out of the FATA and maintain a sizable headquarters in Quetta, Pakistan.[27] Neither they nor their

---

[25] Galula asserts that the DSE transitioned to phase three prematurely.

[26] Murray quotes army troop strength over the course of the war, beginning at approximately 90,000 in 1946 and ending with nearly 200,000 in 1949. Several authors allow that the DSE fielded 20,000 conventional, light-infantry troops at the end of the war.

[27] Some argue that, in addition to several non-Taliban insurgent groups, there are actually several Taliban movements. Kilcullen aptly points out that many Afghans are "accidental guerrillas" who do not necessarily believe in a Taliban ideology but may fight, even temporarily, under the Taliban banner to defend their honor or their property, or for money. One might draw a distinction between the "big T" Taliban in Quetta and elements of the "little T" taliban in rural Afghanistan.

Table 3.2
Number of Insurgencies by Type of Sanctuary

Outcome	No Sanctuary	Involuntary Sanctuary	Voluntary Sanctuary
Government wins	13	3	12
Mixed outcome	5	4	10
Government loses	3	3	20
Ongoing	8	4	4

supporters respect the international boundary.[28] Beyond sanctuary, the Taliban and their al Qaeda allies have set up shadow governments in both areas. Some media reports indicate that elements within the Pakistani government continued to provide the Taliban with financial and logistical support circa mid-2009, even as the Pakistani army units fought the insurgents in semiconventional battles. While the Pakistani civilian government may not support the Taliban, for all intents and purposes, the group enjoys voluntary, if only locally supported, sanctuary. Kilcullen (2009, p. 233) describes the situation on the border this way: "Movement in and out of the FATA . . . is relatively easy, and life can be comfortable and pleasant. Most of the area is a no-go zone for government forces."

When combined U.S. and Northern Alliance forces decimated the Taliban military infrastructure in 2001, the remnants of the Taliban fell back across the Durand Line into Quetta and the FATA to recuperate. By 2006, they had recovered sufficiently to seize control of several districts, if not entire provinces, in southern Afghanistan.

Mullah Mohammad Omar, the Quetta Taliban leader, and most of his lieutenants continued to operate out of the sanctuaries as of late 2009. When Taliban leaders are killed or captured, they are quickly replaced. When leadership losses have an effect on the Taliban, the result is typically tactical rather than strategic. Sanctuary alone cannot keep the Taliban afloat. Other critical factors certainly affect their sur-

---

[28] The border is also commonly referred to as the Durand Line, established in 1893 in an agreement between agents of the British Empire and the then-emir of Afghanistan.

vival and success. The Taliban rely on strong grassroots support from a Pashtun community that feels alienated from both the Afghan and Pakistani governments, on some rural Afghans who have learned to respect the group's toned-down religious message, and on drug-related income and other external support.

If we look at the Taliban insurgency in light of our findings, a rather stark picture emerges. Based on decided cases, voluntary sanctuary theoretically gives them an unweighted 2:1 advantage, and external support a slightly greater than 1:1 advantage. The Afghan government is receiving direct external support, a 5:4 advantage for the Taliban. The Taliban has learned to discriminate in its use of terror, statistically reducing its exposure to popular backlash and shifting its odds (with this correlative factor only) from 5:11 to 14:8. The Taliban is fighting an anocracy in the Afghan government, a 6:1 Taliban advantage. No matter how one interprets the insurgent organization—hierarchical or mixed—the outcome is quantitatively null.[29] And, finally, the Taliban is operating in one of the most impoverished rural areas in the world. Both low income and low urbanization imply an advantage greater than 2:1 for the insurgent.

Since these "advantages" are unweighted, in that we cannot determine that one factor is more strongly correlative than any other, we are not able to determine the relative value of sanctuary for this case (or as a rule) through quantitative analysis. However, a subjective look at the problem suggests that the loss of the Pakistan sanctuary would push the Taliban into semi-isolated internal bases primarily in southern and eastern Afghanistan. This would reduce their capacity to recuperate during the winter off-season; they would find themselves cut off from one of their principal logistics hubs; Taliban cadres would be physically removed from their prime recruiting and training grounds; and grassroots political and logistics support provided by coethnic and

---

[29] We believe the organization to be mixed, with a strong hierarchy representing a central, Quetta-based Pashtu core and with some competing or outlying organizations in the FATA and Baluch areas. Tribal divisions within the Pashtun community create further complexities.

tribal populations would essentially be halved.[30] We found the list of disadvantages posed by the loss of sanctuary to be both extensive and telling, not only for this notional Afghanistan scenario but in nearly every other applicable case as well.

Sanctuary—whether lost or held—is closely correlated with some uncharacteristically dramatic insurgent endings and a number of all-too-characteristic government defeats (see Table 3.2).

Outside Intervention in Support of Government

In 30 of the 89 insurgencies, it was possible to identify significant outside intervention on the side of the government. These interventions occurred 21 times directly (i.e., with ground troops or direct air support) and eight times indirectly (i.e., with money or advisers) (see Figure 3.7).

The results are somewhat counterintuitive. A beleaguered government that gets direct assistance from an outside intervener has no

Figure 3.7
Outside Intervention in Support of Government

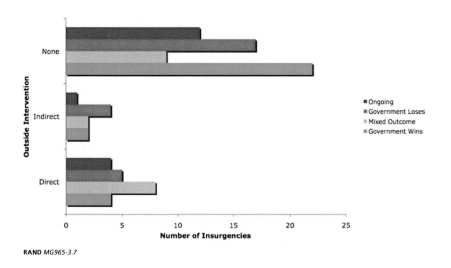

[30] In the absence of an active COIN campaign in southern and eastern Afghanistan, the damage caused by the loss of sanctuary in Pakistan would be nullified as the Taliban simply moved back into Kandahar and other major population centers.

better win-loss record than one that gets no significant assistance. The major difference seems to be that outside intervention is correlated with a higher likelihood of a mixed settlement (specifically, Tajikistan, Lebanon, El Salvador, the Dominican Republic, Bosnia, Cambodia in 1978, and Congo in 1998). The win-loss record for indirect intervention is taken from a very narrow set of case studies, but the finding here also points to the many challenges associated with intervention.

This study generally concludes that governments have a number of advantages against insurgents. It would therefore be logical to assume that the introduction of additional resources would, in many cases, create a tipping point. So why does outside intervention not confer greater benefit to the counterinsurgent? Qualitative analysis of the 30 cases of intervention points to the challenges in achieving just the right balance of support and in providing that support at just the right time. In practice, the quality and consistency of the support provided to the government in each of these cases varied considerably. Even if the supporting government or nonstate actor manages to find that balance, other factors may still portend failure:

- Providing too little or the wrong type of support risks failure, while providing too much risks creation of a weak, dependent state security apparatus.
- Timing of the intervention is critical. Step in too soon and the government can lose credibility or the insurgents might seek sanctuary before they can be engaged; step in too late and the insurgents may already have effected a tipping point.
- Just as the withdrawal of external support to an insurgency can cripple the insurgents, inopportune withdrawal of support for a government can, and very often does, lead to defeat or at least a mixed settlement.
- The behavior of intervening troops can sour the government's information operations or, in some cases, further escalate the conflict. While external influence can speed progress on social reforms that address the root causes of the insurgency, this influence can sometimes be misguided and disruptive. We address this concept further when we discuss anocracies later in this chapter.

- Intervening nations, particularly democracies, are beholden in some way to their citizens. They must sustain popular support or risk being forced into withdrawal.
- A tipping point may have already been reached at the point of intervention, exacerbating the challenges faced by the external sponsor.

Most of our recognized experts mention the need for external support to the government at some point along the conflict timeline. As McCuen (1966, p. 66) puts it,

> [T]he massive mobilization of resources necessary to defeat established revolutionaries in a protracted war presents requirements that few countries can fill from their own means. . . . For example, outside troops probably will be required to defeat a revolutionary movement which has reached the mobile warfare stage because the government power is unlikely to find sufficient, reliable troops to fight against both the rebel regulars and their local forces.

Two cases of direct intervention (North Yemen and Vietnam) and one ongoing case of indirect intervention (Colombia) are instructive.

Yemen, Egypt, and Saudi Arabia. The Egyptian intervention on behalf of the republican government in North Yemen (Yemen 1962–1970) led to a rather muddled settlement. In a desire to further his pan-Arab agenda, President Gamal Abdel Nasser of Egypt helped orchestrate a republican coup d'état that overthrew the North Yemeni royal government in 1962. Nasser then quickly intervened with significant military force and financial assistance to bolster the republicans and, possibly, to establish a base of operations from which he could eventually seize the entire Arabian peninsula. In the face of a combined republican and Egyptian assault, the royalists fled and established an insurgency. In short order, they began seeking the support of neighboring Saudi Arabia. In the first six months of the war, the Saudis remained aloof, resisting royalist pleas for military aid despite harboring their own fears regarding Nasser's ultimate intentions (Badeeb, 1986).[31]

As the COIN campaign against the royalists-cum-insurgents wore on, Egyptian troops increasingly took the lead from the republican Yemeni army. The Egyptian Air Force conducted strikes across the Saudi Arabian border in an effort to draw out the Saudi military. With their villages under attack, the Saudis reluctantly entered the fray. In April 1963, six months after the September 26, 1962, coup d'état, Saudi Arabia began providing political and financial support to the royalists (see Figure 3.8). What appeared at first to be an internecine conflict between two Yemeni factions had now evolved into a proxy war between two external states (Badeeb, 1986).

The North Yemen case reveals lessons for external intervention on behalf of both the government and the insurgents. Egypt fully committed to supporting the republican government, sending tanks, air-

Figure 3.8
End of North Yemen Conflict: Egyptian Perspective

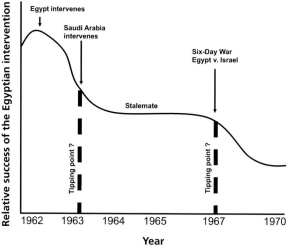

---

[31] Badeeb capably explains the North Yemen conflict in 110 pages of body text. He sources a significant amount of original material.

craft, and, ultimately, at least 40,000 soldiers (Badeeb, 1986, p. 37). Nasser's diplomats and spies encroached on Yemeni sovereignty well before September 26, 1962, and, in all likelihood, Nasser had military forces en route to the Yemeni port of Hudaydah even before the coup plotters took to the streets (Badeeb, 1986, pp. 35–36).

Saudi assistance in support of the insurgents came late and never took the form of direct military assistance. Saudi troops reportedly never crossed into Yemen in force, and the Saudi military provided no substantial equipment or training to the beleaguered Yemeni royalists.[32] However, Saudi financial and moral support was sufficient to give the royalists the temporary boost in power they needed to bring the conflict to a standstill, just a year after the coup (Badeeb, 1986, pp. 37, 55). In 1967, Nasser withdrew a great number of troops to reinforce the Egyptian army against the Israelis.[33] Subsequently, the royalists were able to negotiate a settlement that allowed them to share power with, and participate in, the republican government. Negotiations ending insurgencies are often costly and drawn-out affairs, and this case was no exception.

Arguably, the insurgency ended with neither the republicans nor royalists achieving a clear victory. The Egyptians gained an ostensible ally in North Yemen (Badeeb, 1986, pp. 85–86)[34] but lost more than 25,000 troops and any legitimate hope of controlling the Arabian peninsula. Identification of the tipping point in this case is problematic. One could argue with equal conviction that it came with the beginning of the stalemate in 1963 or as the Egyptian troops withdrew in 1967. Assuming that it occurred in 1963, between 1963 and 1970,

---

[32] The presence of increased Saudi military force along the Yemeni border probably had a positive impact on royalist morale and may have caused Egyptian forces to change tactics or to conduct a brief operational pause.

[33] Fighting continued after the departure of the Egyptian troops until 1970, when the peace plan was fully implemented.

[34] Even this achievement is debatable. Abdullah al-Sallal, the leader of the republican coup and Nasser's ally, was in turn overthrown in 1967. The subsequent administration was much more amenable to reconciliation and, presumably, less likely to support Nasser's pan-Arab agenda in the wake of an eight-year military quagmire.

the Egyptians wasted thousands of lives, years of political capital, and extraordinary sums of money on a lost cause.

As insurgencies go, this one fell short of the median, at eight years. The ending came after seven years of stalemate during which both sides suffered casualties and a loss of political prestige. The negotiations required to finalize the peace plan took three years and required the intervention of the United Nations and several regional state actors. The "tail" of this insurgency was probably the most costly period of the conflict.

Not knowing the outcome of the North Yemen conflict, one might have assumed that the direct and substantial Egyptian support would have given the republican government the political capital and combat power necessary to crush the rather feeble insurgency. Instead, the Egyptians failed to defeat the royalists, became bogged down, and eventually had to withdraw having achieved, at best, a Pyrrhic victory and, at worst, a strategic defeat. The Vietnam War offers another insight into the counterintuitive history of direct external intervention.

Vietnam and the Vietcong. Although the United States had been supporting the GVN since the 1950s, the U.S. military intervention in South Vietnam began in earnest in 1965 with the introduction of major combat units.[35] In 1968, the year in which the total number of U.S. forces peaked at more than 530,000 (Kane, 2006),[36] the Vietcong executed the countrywide Tet Offensive. In response, the Americans and South Vietnamese decimated the Vietcong infrastructure and probably broke the back of the insurgency through a combination of special programs (e.g., civil operations and revolutionary development support, or CORDS) and improved COIN operations. Despite this tactical success, the audacity of the offensive in turn broke U.S. political willpower.[37]

---

[35] South Vietnam is alternatively referred to as the GVN or the Republic of Vietnam in various texts. North Vietnam is referred to by its legitimate pre-1957 title, the Democratic Republic of Vietnam, or DRV.

[36] John Nagl (2005, p. 173) also offers troop counts.

[37] For example, a 1988 article in the *New York Times* ("Tet Offensive," 1988) cites a March 1968 Gallup Poll that showed a sharp decline in support for the war in the wake of President

Within three years of the Tet Offensive, the Vietcong were relegated to a supporting effort.[38] North Vietnamese army units engaged U.S. and South Vietnamese forces in what became increasingly conventional combat operations, and the conflict had almost fully given over to conventional, or Maoist final-phase (Mao, 1961 [2000], pp. 42–43), war by 1972.[39] In the same year, U.S. troop levels had been reduced to just over 35,000, a number signifying a marked decline in direct military support to the GVN. In 1973, the United States began to withdraw even indirect financial support for the South Vietnamese, and North Vietnamese conventional forces struck the final blow in 1975.[40]

Although the GVN may have been facing an eventual long-term defeat, some analysts believe that the 1973 Paris Peace Accords could have forestalled this eventuality. The reductions in financial support between 1973 and 1975, however, severely undermined the peace accords, as well as the South Vietnamese morale and economy.[41] The withdrawal of financial aid, an act stemming in no small part from the moral victory of the Vietcong in 1968, was a significant causal factor in collapse of the South Vietnamese government (Tower, 1981–1982).[42]

Johnson's failure to convince the American public of what was, in many respects, a battlefield victory. Most of our COIN experts concur (e.g., Beckett, 2001, p. 190).

[38] Jeffrey Record (2007, p. 49) describes the insurgency as "effectively contained" by 1971.

[39] This is often referred to as the *third phase* of a Maoist insurgency. One can interpret Mao as having identified seven phases of conflict, but conventional wisdom identifies three.

[40] Loss of U.S. logistics support was especially crippling to the Army of the Republic of Vietnam (ARVN) and the Republic of Vietnam (RVN) Air Force. By late 1974, one-fifth of the air force was grounded for lack of fuel and spare parts, artillery batteries were rationed to three rounds per tube, arms stocks were depleted by 25 percent, and infantrymen were rationed to 85 rounds per day (Clodfelter, 1995, p. 207).

[41] The Paris talks were held in large part due to increases in U.S. bombing operations and the setback of the 1972 Easter Offensive, during which the North Vietnamese Army (NVA) may have lost 100,000 troops. Qiang Zhai, a recognized expert on People's Republic of China (PRC) intervention in Vietnam, also states that the Chinese urged the North Vietnamese to the peace table in order to give President Nixon a face-saving vehicle for withdrawal. PRC leadership had full intent on pressing for forced reunification of Vietnam one to two years later (Zhai, 2000, pp. 202–215).

[42] The Fulbright amendment to the fiscal year 1974 appropriations act ended any hope of enforcing the Paris Peace Accords (Wiest, 2002, p. 80). When the United States cut funding

In this case, as in many cases of failed intervention, the state collapsed very quickly once the insurgents (or, here, the NVA) gained a clear upper hand. The precipitous nature of this collapse—tanks rolling through the streets of the capital—comes into greater contrast when one considers the lengthy U.S. involvement in the postcolonial and advisory periods leading up to 1965.

Despite some ongoing tactical and operational successes— particularly against the Vietcong—a tipping point had been reached in 1968. U.S. willpower flagged as quickly as, or perhaps more quickly than, the Vietcong insurgency. The operational defeat of the insurgent cadre in this case proved irrelevant, since the sponsor abandoned its overall strategic campaign at the height of insurgent violence. It is unlikely that, even with continued support, the GVN would have lasted more than a few years against an increasingly powerful and professional NVA. In the end, half a million U.S. troops, thousands of tons of ordnance, and billions of dollars in national treasure all failed to secure the fate of South Vietnam (see Figure 3.9).

This somewhat narrow insight into the end of the Vietnam War does not touch on all the contributing factors that led to North Vietnamese victory, including the quality of the Vietcong and NVA forces, the inability of the U.S. military and policymakers to grasp the central tenets of COIN theory until late in the war, the availability of sanctuary in neighboring Laos and Cambodia, the anocratic nature of the GVN, operational failures by the ARVN in the early 1970s, the brutally effective Vietcong indoctrination program, or the logistical support provided to the North by the Soviet Union and the PRC. It does, however, accurately reflect the quantitative findings. North Yemen, Vietnam, and the other cases of failed or faltering direct interventions (notably, Afghanistan 1978–1992) all show the pitfalls and often overwhelming complexities inherent in such operations.

It seems that interventions end badly more often than not. Successful interventions, when they occur, tend to follow the average trend

---

in half between 1973 and 1974, "The drop in aid caused the South Vietnamese economy to implode . . . causing widespread poverty and unrest. The South Vietnamese people had little reason to defend their nation."

Figure 3.9
Fall of the Government of Vietnam

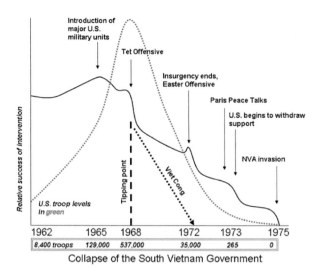

Collapse of the South Vietnam Government

to end state. A few take longer to achieve effect, and there are myriad differences between direct and indirect intervention. We provided a narrow glimpse into direct interventions in North Yemen and Vietnam. The case of U.S. support for the government of Colombia (1963–present) offers an example of what may come to be a successful indirect intervention.

Colombia and the Fuerzas Armadas Revolucionarias de Colombia, or Revolutionary Armed Forces of Colombia. Although we judged it to be "ongoing," the COIN campaign against the Revolutionary Armed Forces of Colombia (FARC) has been meeting with increasing success since 2000 (see Figure 3.10). From its inception through the early 1980s, the FARC insurgency opposing the Republic of Colombia operated in remote areas of the country and was unsuccessful in gener-

Figure 3.10
Rise and Fall of the Fuerzas Armadas Revolucionarias de Colombia, 1963–2009

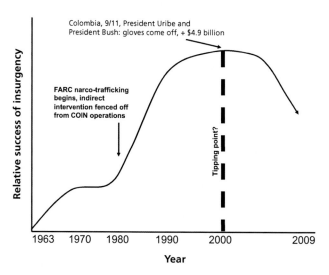

RAND *MG965-3.10*

ating a significant threat to the government.[43] Equally, the government proved incapable of building the political will or military strength necessary to crush the insurgents. Until the FARC actively engaged in drug trafficking circa 1981, the Colombian COIN campaign received little international attention (Rabasa, Warner, et al., 2007, p. 67).

President Ronald Reagan's escalation of the war on drugs led to indirect U.S. intervention in Colombia.[44] From 1981 to 2000, U.S. support trickled or flowed into the country, with fluctuating rates and values depending on national budget issues, presidential focus, and congressional sentiment. Almost all of the military support was fenced off solely to support counternarcotics operations, and, throughout the

---

[43] Marc Chernick (2007, p. 52), a noted FARC scholar, asserts that the current violence is rooted in the 1940s and 1950s. We do not dispute this point; we have focused our case examination specifically on the 1960s–2009 FARC.

[44] President Nixon first used the term *war on drugs* in 1971, concurrently establishing the Special Action Office for Drug Abuse Prevention (SAODAP) (Public Broadcasting Service, undated).

late 1980s and all of the 1990s, the Colombians did not capitalize on this external, indirect intervention. Meanwhile, the FARC grew stronger, obtaining more and advanced weaponry. The insurgents eventually took control of a large autonomous zone within Colombia's borders (Public Broadcasting Service, undated). Between 1993 and 2000, coca cultivation reportedly had increased by more than 300 percent, and opium-poppy cultivation reportedly had increased by 75 percent. Approximately 3,000 Colombians were reportedly kidnapped in the year 2000 (GAO, 2008, p. 9).[45] At this point, the insurgency was just shy of the 40-year mark with no end in sight.

In 2000, President Clinton announced the details of U.S. participation in Plan Colombia, a program developed jointly with then-President Andres Pastrana (GAO, 2008, p. 11). The events of 9/11 and the focus on defeating global terrorist networks gave the program a critical jump-start. The U.S. support package would eventually include at least US$4.9 billion in military support (GAO, 2008, p. 5).[46] Although the U.S. Congress altered Plan Colombia to a certain extent, the flow of indirect support to the Colombians increased significantly. By late 2002, the Plan Colombia money pledged in 2000 began to show an impact.

By 2009, 46 years into the insurgency,[47] the Colombians had achieved several ringing military victories, had reduced the FARC area of control, conducted a dramatic rescue of FARC-held hostages, and managed to turn public and international opinion against the insurgents ("Hostages Held for More Than Five Years Rescued in Colombia," 2008). The Colombian military reported a near 25-percent nationwide drop in homicides, a 90-percent drop in kidnappings, and a 20-percent reduction in the size of FARC-controlled territory between 2000 and 2007 ("Hostages Held for More Than Five Years Rescued in Colombia," 2008, pp. 22–25, citing Colombian government figures). Most telling, though, are the reports of desertions and defections and the

---

[45] By comparison, one source cites a reported 243 kidnappings in Nigeria, a country three times as populous as Colombia, in the same year. See Cleen Foundation (undated).

[46] This number may be as high as US$8 billion or 9 billion as of 2009.

[47] Marc Chernick (2007, p. 55) places the formal establishment of the FARC in 1966.

increased reporting provided to the military by Colombian citizens (see Figure 3.11).

Between 2000 and 2009, the rates of defection and desertion increased significantly. The Colombian government reported a greater-than-50-percent reduction in FARC membership (17,000 to 8,000) between 2001 and 2007 ("Hostages Held for More Than Five Years Rescued in Colombia," 2008, p. 25). The rate of defections alone increased from 2007 to 2008 by a further 20 percent (Kraul, 2009). Another article excerpt quotes the Colombian defense minister:

> Guerrillas have been deserting in droves: since the beginning of 2007, more than 3,900 have turned themselves in to the authorities, according to the defense ministry. Juan Manuel Santos, the defense minister, says that in recent months a bigger proportion of the renegades are experienced fighters with detailed knowledge of military operations. "A couple of years ago the people who left had been in the FARC for only a few years. Now on average they have served between 15 and 20 years," he says. (Lapper, 2008)

Figure 3.11
Tipping Point of the Colombian Insurgency

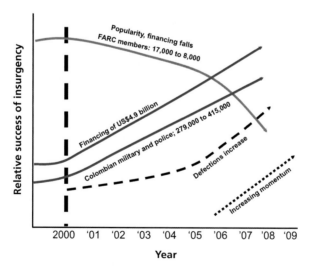

In 2008, a group of more than 1,000 formerly steadfast FARC prisoners formed a defector group called "Hands for Peace." Members of this group are offered amnesty in return for providing information on criminal activities and for renouncing their allegiance to the FARC (Forero, 2008). The anecdotal evidence showing the impact of FARC defections is nearly as compelling as the government statistics. Keeping in mind the cautionary tale of the Chieu Hoi program, however, propaganda and statistics should be analyzed with caution. In the case of Vietnam circa 1968, tactical momentum was counterbalanced by a crushing moral defeat. The GVN was incapable of consolidating control of its territories, did not institute acceptable democratic reforms, and proved overly dependent on direct U.S. combat support (troops, equipment, firepower). What appeared to be "positive" end-state indicators proved misleading.

In contrast, the Colombian government seems to be building momentum. President Uribe has sustained the legitimacy of the Colombian democratic process, he has strengthened the army and police forces to the point that they can probably stand without (or with less) U.S. aid, and he has successfully rallied the majority of the Colombian people to his cause. Taking all these factors into account, defections by senior FARC insurgents in 2007 and 2008 eventually may prove to be as significant as the officer defections that signaled the tipping point in the Cuban revolution in 1958. In the case of Colombia, indirect U.S. support (small adviser deployments, financing, intelligence, and arms sales) may tip the balance in favor of the Colombian government.[48]

For a variety of reasons, the United States was able to successfully adapt its support program to reverse a faltering COIN campaign in Colombia. Similarly, as its COIN efforts against the Chinese Malayan insurgents dragged on unsuccessfully, Britain modified and increased its operational tempo to achieve victory. A change in U.S. force posture

---

[48] As of mid-April 2009, criminal groups and other armed gangs were already starting to step up acts of violence and intimidation in order to wrest control of some narco-trafficking business from the more prominent insurgent groups. In the absence of a paradigm shift in drug consumption in the United States, elements of the FARC will almost certainly morph into purely criminal narco-traffickers.

and tactics in Iraq in 2007 and 2008 seems to have salvaged (at least temporarily) an operation that had been widely written off as a strategic defeat. A qualitative overview of the remaining cases shows that a willingness and capacity to modify strategy in the midst of a campaign can prove decisive. Conversely, Egypt's failure to modify its methods in North Yemen correlated with mixed outcome or defeat, depending on interpretation of the data.

At its outset, this study posed a series of questions that a policy-maker might ask: Is the prospective COIN operation viable? Is it worth the anticipated risk to international prestige and treasure? Do conditions on the ground seem to favor a successful COIN campaign, or do they suggest probable failure? A quantitative finding that shows intervention on behalf of the state to be a less than 50-50 proposition begins to answer these questions (see Table 3.3).[49]

Outside Intervention in Support of Insurgencies

State-supported insurgencies in our data set have won more than half of the time (see Figure 3.12). Those with nonstate support have won and lost at about an equal rate, and those with no outside support whatsoever have won only three of 18 times. Where state support started but was cut off, insurgents did far worse than average.

Qualitative analysis of the 52 cases in which states supported the insurgents shows that consistent state support is most critical as the conflict tips toward end state. Insurgents generally concur; Mao

Table 3.3
Number of Insurgencies by Outside Intervention

Outcome	Direct	Indirect	None
Government wins	4	2	22
Mixed outcome	8	2	9
Government loses	5	4	17
Ongoing	4	1	12

---

[49] This ratio takes into account government wins and government losses only.

Figure 3.12
Level of State Support to Insurgency

NOTE: The number of insurgencies in each category was normalized to 28 (the number of insurgencies in the government-won category) by arraying insurgency lengths in monotonic order and then inserting interpolated results evenly throughout the range. This method does not significantly change the percentile rankings of the results.
RAND *MG965-3.12*

(Mao, 1956, p. 173, as cited in McCuen, 1966, p. 64) puts it as follows: "International support is necessary for the revolutionary struggle today in any country or of any nation."

We discovered a relatively robust data set from which to examine external support to insurgencies. Insurgents received some level of support in 75 percent of all cases, and, of the remaining cases, only one insurgency succeeded. In and of itself, this statistic reveals the criticality of external support to the insurgent. O'Neill (1990, p. 111) summarizes conventional wisdom of the counterinsurgents on this point:[50]

> Unless governments are utterly incompetent, devoid of political will, and lacking resources, insurgent organizations normally must obtain outside assistance if they are to succeed. Even when substantial popular support for the insurgencies is forthcom-

---

[50] Insurgent theorists tend to gloss over the standard notion of external support, focusing instead on "international" or "continental" revolutions that are mutually supporting. In other words, insurgencies support insurgencies, both morally and physically.

ing, the ability to effectively combat government military forces usually requires various kinds of outside help, largely because beleaguered governments are themselves beneficiaries of external assistance.

The manner and consistency of this support often determines how the insurgency ends. Support provided by diasporas, nonstate actors, and states range from informal financial aid to direct military intervention. Galula (1964 [2006], p. 39) breaks this down further into moral, political, technical, financial, and military support. O'Neill substitutes "material" for technical, financial, and military, but he is generally in line with Galula. In some cases, this support—however it is defined—remains consistent in form, quality, and quantity over time. Strong and relatively consistent Chinese and Soviet intervention in Vietnam clearly played a substantial role in ending the war in favor of the North.[51] The Tamil and Irish-American diaspora provided moderate levels of financial support to the LTTE and the PIRA, respectively, although it would be difficult to argue that this support helped the insurgents shape a positive end state. Table 3.4 offers an additional take on the various types of external support available to insurgents. The table, built in support of a 2001 RAND study titled *Trends in Outside Support for Insurgent Movements* (Byman, Chalk, et al., 2001, p. 105), shows the emphasis sponsors historically have placed on various types of support.

External governments are by far the most aggressive sponsors of insurgencies. The relative value of this external support over time is a critical factor in shaping end state, especially when compared to other correlative factors. By juxtaposing Chinese intervention with the timeline for the collapse of the South Vietnamese government, we can show a correlation between declining U.S. support and steadier Chinese support for the Vietcong and NVA between 1962 and 1975.[52] Concurrent

---

[51] Over the course of the war, hundreds of thousands of Chinese troops (possibly 150,000 at peak in 1967) and thousands of Soviet advisers and antiaircraft specialists were sent to North Vietnam. Both countries also sent thousands of tons of equipment and supplies.

[52] This does not take into account significant Soviet support to the DRV. Citing Lanning and Cragg's (1992) *Inside the VC and NVA,* Record (2007, p. 52) provides detail on Soviet

Table 3.4
Sources of Contributions

Form of Support	State	Diaspora	Refugees	Other Nonstate
Money	Significant	Significant		Minor
Safe haven	Significant		Minor	Minor
Diplomatic backing	Significant	Limited		
Arms	Significant	Minor		Limited
Training	Significant	Minor		Minor
Intelligence		Minor		
Direct military support	Significant		Limited	Minor
Inspiration	Limited	Minor		Minor

SOURCE: Byman, Chalk, et al. (2001, p. 105).

with the sharp drawdown of U.S. forces in 1968–1969, Chinese commitment to the North began to fade. However, the PRC continued to provide significant material, financial, and other aid to the North nearly to the end of the war—comparatively quite a bit more than the United States was, at the time, providing to South Vietnam. The PRC provided a bump in support in 1972 to replenish NVA equipment lost in the Easter Offensive just as the U.S. Congress was beginning to consider the cessation of financial and military support to the GVN.[53]

---

support between 1965 and 1972: 340 aircraft, 711 pieces of artillery, 132,000 cases of artillery ammo, 90 surface-to-air missile (SAM) batteries, 165 radars, and 131,000 small arms. Later, they sent additional arms and munitions, including 400 T-54 tanks and thousands of trucks.

[53] The red line in Figure 3.13 is a notional representation of both Chinese troop levels and aid. Figures are derived from a compilation of sources including Li (2007, pp. 217–219), Record (2007, pp. 50–51), and Zhai (2000). Zhai references several original Chinese sources, including Xie Yixian. Quoting Yixian, he states that, "in 1969 Beijing provided Hanoi with 139,900 rifles, 3906 pieces of artillery, 119.2 million bullets, and 1.36 million artillery shells, as compared to 219,899 rifles, 7087 pieces of artillery, 249.7 million bullets, and 2 million artillery shells in 1968" (Zhai, 2000, pp. 179–180). Jeffrey Record lists Chinese aid from 1964 through 1975: 1,923,000 small arms, 64,530 artillery and mortar

Figure 3.13 infers only a correlation, not sole causation in the collapse of the GVN.[54] However, Record believes that external support was, if not solely causal, clearly critical to the success of North Vietnam:

> [It is] impossible to argue that the Communist victory in South Vietnam was anything other than a triumph of foreign help. Indeed, the Communists could not have fought the war or won it the way they did without massive support from China and the Soviet Union. . . . External assistance is a common enabler of vic-

Figure 3.13
Effects of External Support, Vietnam

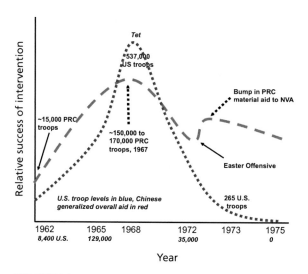

RAND MG965-3.13

---

pieces, 1,707,000 artillery shells, 30,300 radios, 560 tanks, 150 naval vessels, 160 aircraft, 15,770 wheeled vehicles, and 11,400,000 sets of uniforms.

[54] In 1973, relations between the Chinese and the North Vietnamese soured when preexisting territorial and regional disputes resurfaced and as relations between China and the Soviet Union soured. Although he does not infer that China decreased its material or financial aid to North Vietnam after the Easter Offensive bump, Zhai makes it clear that Chinese enthusiasm for the Vietnam War was beginning to flag at this point. The red line in Figure 3.13 reflects this drop in moral support between 1973 and 1975. By this time, however, it was too late for the GVN (Zhai, 2000, pp. 210–211).

torious insurgent wars, though certainly no guarantee of victory. (Record, 2007, pp. 48, 133)

We concur with Record in that, at best, North Vietnam would have been able to sustain an ongoing and perhaps low-level insurgency in the absence of Chinese and Soviet aid. A brief look at the cited data shows the sheer magnitude of combat power delivered by these external sponsors to the North Vietnamese, allowing one to imagine a different outcome in its absence.

Clear-cut and sustained support, then, seems to be a critical requirement for insurgents. What about those cases in which the consistency of external support changed significantly over time? How did those changes affect the outcome of the conflict?

Angola and UNITA. The North Yemen insurgency showcased external support for both government and insurgent forces. Jonas Savimbi's União Nacional para a Independência Total de Angola (UNITA, or National Union for the Total Independence of Angola) insurgency against first the Portuguese colonials and then the Movimento Popular de Libertação de Angola (MPLA, or Popular Movement for the Liberation of Angola)[55] provides an excellent opportunity to explore the dynamics of unsustained external support.[56] By the time Savimbi was killed in an ambush in 2002, UNITA had received financial or military aid from at least a dozen foreign governments, including the United States. None of these interventions was sufficient to carry Savimbi to victory. The UNITA insurgency ended in a series of tactical and diplomatic defeats that played out in a "tail" that lasted from 1992 through 2002.

In the mid-1960s through the early 1970s, several Angolan insurgent groups fought to overthrow the Portuguese colonial government. Savimbi, an anticolonial insurgent, formed UNITA in 1966 after a 1964 split with the larger Frente Nacional de Libertação de Angola

---

[55] MPLA started alongside UNITA as an anticolonial insurgency before assuming control of postcolonial Angola in 1975.

[56] Quantitative data for this study covered the UNITA insurgency from 1975 through 1992. Pre-1975 narrative is provided for illustrative purposes.

(FNLA, or National Liberation Front of Angola) insurgent group.[57] Having been briefly trained in the PRC, Savimbi was favored to receive PRC sponsorship.[58] The Chinese, in an effort to expand their foothold in Africa and, more importantly, to undermine western and Soviet global interests, provided UNITA a few weapons as well as training for a small pool of insurgents.[59] However, the PRC never fully committed to supporting UNITA. At no point during the period of indirect intervention did China deploy a sizable advisory or training team to shape the UNITA fighters into competent insurgent cadres.

In the absence of dedicated support, Savimbi made an effort to follow the Chinese Maoist model of insurgent warfare: He practiced self-sufficiency. Poorly armed and poorly trained, UNITA made several daring yet uncoordinated raids against Portuguese targets in 1966. The Portuguese repulsed the first wave of attacks, inflicting losses on UNITA columns.[60] China's half-hearted intervention on behalf of UNITA in the mid-1960s did more to place the fledgling insurgency at risk than to further China's national agenda in Africa.[61] In the long run, the Chinese would prove to be fickle sponsors, switching support

---

[57] Savimbi informally established UNITA in 1964 and then "officially" in 1966 (Potgieter, 2000, p. 256).

[58] Various sources state that Savimbi and 11 UNITA insurgents were trained at either the "Nanking Military Academy" or "Political Warfare Academy (in Nanking)" in 1964 (e.g., Bridgland, 2002). The article, with a Johannesburg dateline, quotes MPLA's ambassador to Great Britain circa 2002. Coincidentally, Castro and Guevara also started out with 11 men in Cuba.

[59] Zhai (2000, pp. 140–143) describes in detail the paranoia that led Chairman Mao Tse-tung to believe that a U.S. or Soviet conventional attack against China was not only possible but probable. This belief drove Chinese interventionist policies from the mid-1960s through the early 1970s. Mao's belief in self-reliance caused him to either lessen or withdraw support for several insurgencies midstream. This happened in the cases of Vietnam circa 1969 and Angola circa 1966.

[60] They eventually succeeded at the battle of Teixeira de Sousa in interdicting international rail lines, a tactic that backfired. The temporary loss of rail traffic stopped commerce from reaching Zambia, a key UNITA sponsor.

[61] It is entirely possible this was China's objective. Not all external sponsors have the best interests of the insurgency in mind.

from one insurgent group to another as various fortunes ebbed and flowed.[62]

By many reports, Savimbi was a competent leader with sound organizational skills. If the Chinese had committed to full indirect intervention in support of UNITA through 1966, it is possible, if not likely, that Savimbi would have been successful in overtaking MPLA and FNLA.[63] He would have been well positioned to take power as the Portuguese buckled in 1974, providing the Chinese with a friendly government in the resource-rich heart of the continent. Prior to 1969, the United States' executive branch paid scant attention to sub-Saharan Africa, and the Johnson administration would have been unlikely to directly oppose the establishment of a PRC puppet regime in Angola (Marcum, 1976, p. 406).

At the zenith of the pan-Arab and Non-Aligned movements,[64] prior to the June 1967 Arab-Israeli War, Gamel Abdel Nasser formally committed Egypt to UNITA's cause. The PRC continued to send small shipments of aid for a few years, but, by 1967, the Egyptians had transplanted China as Savimbi's de facto sponsor (Potgieter, 2000, pp. 256–258). The Egyptians supported UNITA in the late 1960s and continued to provide some indirect financial and military aid throughout the early 1970s. Savimbi is quoted as identifying Egypt as the sole supplier of arms and money to UNITA from the end of 1966 through 1974 (Marcum, 1978, p. 228).[65] Mirroring the ineffectual efforts of the

---

[62] The PRC supported several anticolonial movements in Africa in the late 1960s and, particularly, in the early 1970s. They reportedly sent more than US$2 billion in economic aid to various African nations between 1954 and 1975. However, the PRC tended to back losing causes. None of the PRC-backed insurgencies in Mozambique, South Africa, Namibia, Zimbabwe-Rhodesia, or Angola bore fruit, in great part because the Chinese were only just building efficiency in supporting international socialist movements. They often found themselves outpaced by the Soviets and eventually shifted policy to support some Soviet-backed insurgent groups (Klinghoffer, 1980, pp. 101–102).

[63] UNITA both collaborated with and competed against the FNLA and MPLA throughout the course of the conflict.

[64] Nasser was also a cofounder of the Non-Aligned Movement of developing, anti-imperial, postcolonial nations.

[65] This was probably an exaggeration.

PRC, however, Nasser failed to provide Savimbi with ample arms shipments or trainers.[66]

Inconsistent support made for uninspiring performance. Prior to 1974, UNITA was the smallest of the three insurgent organizations, fielding between 2,000 and 3,000 fighters in the mid-1970s.[67] The group reportedly was responsible for less than 5 percent of attacks against Portuguese forces in 1970 (Marcum, 1978, p. 217). When the Portuguese colonial government collapsed in 1974, UNITA had to contend with a much stronger MPLA organization fully backed by the Soviet Union.

After the 1974 collapse, Angola became a proxy battlefield on a number of fronts. Fading relations with Egypt and China at first proved irrelevant to Savimbi as the French, the West Germans, South Africans, and even Saudi Arabia backed UNITA and the FNLA against the new Soviet-aligned MPLA government.[68] Small sums of money and equipment began to flow, and, in 1975, South Africa invaded Angola in support of the UNITA-FNLA coalition.[69] In response, the Soviet Union flooded MPLA with aid shipments that included T-34 and T-54 tanks, 122mm rockets, and combat aircraft. Cuba added thousands of troops, tipping the balance in favor of MPLA. South Africa was forced to withdraw in ignominy, and the poorly trained UNITA and FNLA troops pulled back to their tribal cantonments in defeat. External funding to UNITA quickly ebbed (see Figure 3.14) (Marcum, 1976, p. 417).

At the end of 1975, nine years into the insurgency, UNITA found itself in control of a sizable piece of Angolan tribal territory but no closer to seizing the capital, Luanda. The UNITA insurgency eventu-

---

[66] From 1966 through 1974, UNITA, MPLA, and the FNLA entered into and then broke a series of cease-fires and partnerships. When MPLA assumed control of Angola in 1975, UNITA and the FNLA entered into another doomed compact.

[67] Marcum (1978, p. 257) quotes 3,000 and 8,000 from two separate sources, while others, including Klinghoffer (1980, pp. 15, 25), offer 2,000 or 4,000 as reasonable. The other groups all fielded three to four times that number.

[68] Marcum and Klinghoffer provide extensive detail on various aid packages provided to UNITA between 1974 and 1976.

[69] South African motives were complex but revolved around anti–South African insurgents havened in Angola.

Figure 3.14
União Nacional para a Independência Total de Angola Support, 1964–1976

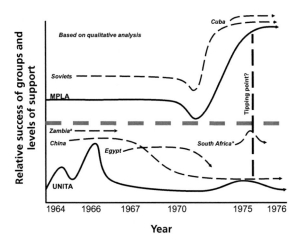

NOTE: MPLA = Movimento Popular de Libertação de Angola, or Popular Movement for the Liberation of Angola. UNITA = União Nacional para a Independência, or National Union for the Total Independence of Angola. The figure does not show support from Tanzania, Zaire, Bulgaria, Morocco, Saudi Arabia, and some other states.
aZambia later renewed support to UNITA.
RAND *MG965-3.14*

ally ended in failure at the 26-year mark. Although Savimbi had managed to obtain the support of a dozen foreign governments, including the United States, he never was able to get any of his sponsors to fully commit. In each case, a sponsor provided UNITA some level of support, only to abandon the cause within a few months or years. UNITA insurgents never benefited from sustained training, never received a steady flow of funds or equipment, and often found themselves simply subsisting (see Figure 3.15). Marcum (1978, p. 221) states,

> The experience of UNITA illustrates how the absence of appreciable external support can limit insurgent capacity. Deprived of a contiguous staging base and unable to obtain substantial material assistance from outside, UNITA's ill-armed guerrillas were unable to capitalize on [strong ethno-political roots].

Figure 3.15
Rise and Fall of União Nacional para a Independência Total de Angola

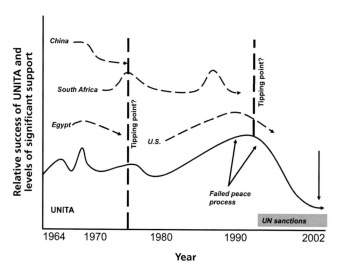

RAND *MG965-3.15*

Record believes that support to insurgents is equally as important as support to the counterinsurgent, stating plainly that the Algerian FLN and ethnic Chinese Malay insurgents were both defeated primarily because they were cut off from aid. He concludes,

> It seems no less reasonable to conclude that highly motivated and skilled insurgents can be defeated if denied access to external assistance and if confronted by a stronger side pursuing a strategy of barbarism (or one that) combines effective grievance redress and discrete use of force. (Record, 2007, p. 57)

While it is not hard to understand why the total withdrawal of support often signals the inevitable collapse of an insurgency, analyzing the effects of limited or sporadic support is more challenging. Proving that inadequate support was the primary causative factor in ending an insurgency may be an exercise in futility. In each of the cases in which limited support was provided, the level and efficacy of that support ebbed and flowed over the course of the insurgency. At certain points in a conflict, the external actor ensured the insurgents' survival, while, at other points, undermining its operations. Further, limited support does not always presage defeat: Limited, indirect Saudi assistance to the North Yemen royalists was sufficient to bring the government to the bargaining table.

In the face of this uncertainty, the UNITA case study offers some insights. The manner in which the PRC supported UNITA is representative of other cases in which limited sponsorship contributed to failure. Savimbi probably had a clear moment of opportunity in 1966, one on which he could not capitalize, in large part, due to a lack of resources and training. A comparatively enfeebled UNITA never had a genuine opportunity to eclipse MPLA before 1975. Savimbi's tipping point may have come in 1966.

It appears that supporting foreign insurgencies is generally imprudent (other than for specific, short-term objectives or a general intent to cause disruption). That said, there are mitigating factors that sponsors may consider. These conclusions are based on the quantitative analysis in this study and reinforce conventional wisdom:

- The external sponsor cannot control most of the causal factors leading to end state. Thorough and persistent intelligence analysis is the only foil against unexpected failure and is a key component in most successful interventions.
- Operational and strategic flexibility have proven crucial in nearly all cases of intervention. Campaign planning should be founded on the expectation that significant shifts in requirements, momentum, and direction are all but inevitable.
- Many of the factors that affect the outcome of interventions on behalf of the government apply to interventions on behalf of the insurgents. Timing is equally critical in both cases. Galula (1964 [2006], p. 42) believes that, "If outside support is too easily obtainable, it can destroy or harm self-reliance in the insurgent

ranks. . . . [O]utside support (should come) in the middle to later stages of the insurgency."[70]

- We found that partial interventions, or those lacking focus and dedicated support, are likely to fail. Sponsors should intervene only when they are fully dedicated to the proposed mission and believe that they will be able to sustain their commitment for ten to 16 years.

This last point assumes that sponsors want the insurgency to succeed and will support the insurgents for the duration of the conflict. O'Neill (1990, p. 119) points out that this is not always the case (see also Table 3.5):

> The fact is, few if any external states engage in open-ended assistance programs for altruistic reasons; they render support because it serves their interests at specific points in time. As a result, it is not unusual to find that they often decrease or terminate assistance or, in some instances, switch sides if it suits their purposes.

Finally, O'Neill (1990, p. 121) describes another scenario in which the sponsor purposefully abuses its relationship with the insurgents:

> (Some) donor states contribute to internecine strife . . . by backing one group at the expense of its rivals. This can result from an

Table 3.5
Number of Insurgencies by Level of State Support

Outcome	State Support (which ended)	Nonstate Actor Support Only	No Support
Government wins	11 (8)	5	12
Mixed outcome	12 (7)	3	4
Government loses	21 (2)	4	1
Ongoing	8 (3)	3	5

---

[70] This is Galula's informed opinion and not a conclusion of *How Insurgencies End*. In some cases, early intervention may be best, especially when addressing protoinsurgencies.

outright desire to establish hegemony over the insurgents, from a perceived need to check the influence of other donor states, or, as is so often the case, from both.

CHAPTER FOUR

Assessments of Insurgency Endings: Internal Factors

Structure

Hierarchical insurgencies are those organized in accordance with military "line and block" charts, establishing relatively clear, vertical chains of command, while networked insurgencies adhere to a flat organizational structure with vague leadership roles for peripheral subgroups (see Figure 4.1).[1] Networked organizational diagrams are horizontal or cloudlike, but often with a small, centralized leadership chain. In terms of structure, we identified that most insurgent groups are hierarchies— as one might expect from a quasi-military organization. With the emergence of networked international terrorism, a presumption has arisen that networking is a useful innovation for insurgents, making them more resilient and flexible, and thereby more likely to win. The numbers do not yet bear out such a belief (see Table 4.1).

While the concept of hierarchy is rather simple, insurgencies rarely are. With a few exceptions, insurgencies operate along both military and political lines and, therefore, field both a military and political wing. In some cases, the political hierarchy may take the form of a legitimate party, while, in others, it may exist as an underground shadow organization closely linked to the insurgent cadre.[2] O'Neill

---

[1] This section primarily addresses military hierarchy but touches briefly on separate political hierarchies (e.g., PIRA and Sinn Fein).

[2] When Velupillai Prabhakaran was alive, he controlled both the political and military wings of the Sri Lankan LTTE (O'Duffy, 2007, p. 265). This type of direct military and political hierarchy to a single prominent figure is not necessarily rare, but it does seem to be

Figure 4.1
Insurgent Structure

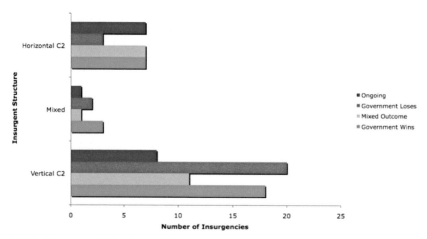

NOTE: C2 = command and control.
RAND *MG965-4.1*

(1990, pp. 91–107) uses the term *parallel hierarchy* to describe this phenomenon.[3]

Networks are perhaps even more complex. The Uruguayan Tupamaros insurgency provided us with the clearest and best-recorded example of a networked insurgency. We will continue to refer to the Tupamaros in this and in the following section, where we will also describe the end of the conflict. This excerpt describing the Tupamaros organization well defines the basic infrastructure of a networked insurgency:

---

rare in long-lasting insurgencies. Councils seem to prove more survivable over time, although our data do not necessarily bear this out.

[3]   In a lengthy but informative footnote, McCuen (1966, p. 98) states,

> This concept of "parallel hierarchies" has become fundamental to French revolutionary war concepts. French theorists consider that their major problem is to parry or destroy the parallel hierarchies of the revolutionaries and then replace them with parallel hierarchies of their own. . . . The concept is basically sound.

McCuen disagrees with himself, however, threading the concept of unified command and unified effort throughout his book.

Table 4.1
Number of Insurgencies by Insurgent Structure

Outcome	Vertical C2	Mixed	Horizontal C2
Government wins	18	3	7
Mixed outcome	11	1	7
Government loses	20	2	3
Ongoing	8	1	7

As their numbers and facilities grew, the Tupamaros developed an elaborate organization designed to maximize fighting potential while insulating the movement against detection and destruction. Following the concept of compartmentalization, the structure consisted of a central executive committee and columns of 30 to 50 people, each containing cells of between 5 and 10 members. Although coordinated from above, each column was designed to be self-sustaining in the case of trouble at the top; thus the columns were equipped to gather intelligence, maintain supplies, and undertake armed or propaganda action independently, making it theoretically possible for one surviving column to regenerate the movement. Outside the underground itself were support committees that helped in recruiting, securing supplies, and providing needed skills. (Wright, 1991, p. 99)

Although we can describe both types of insurgencies, clearly identifying the organizational structure of an insurgency, even in hindsight, is challenging. For many of the 89 cases, one could argue that insurgent organization was both hierarchical and networked with near equal credibility. For example, the Chechen insurgents fighting against the Russians in both the 1994 and 1999 conflicts relied on both a centralized military command structure and a distributed network of militia (Oliker, 2001, pp. 39–41; Kulikov, 2003).[4] Similarly, from 1991 to 1993, Somali tribal leaders relied on their own linear command structures but also on the ability to call individual armed Somalis to the streets. Insurgent structures also change over the course of a campaign, typically with purpose. Maoist insurgent methodology calls for sequential transition from a protoinsurgency to a loosely organized movement to structured guerrilla bands and then to a hierarchical conventional force, with the latter typified by the Colombian FARC and the Greek DSE.[5] Insurgencies can also revert from hierarchy to network as a survival technique. Some insurgencies consist of more than a single organization (e.g., UNITA, FNLA, and MPLA) but do not necessarily form a networked whole.[6]

Statistical evidence shows that networked insurgents have lost significantly more often than they have won, while hierarchically organized insurgents have a more even record. It remains to be determined whether this result reflects the fact that networked organizations do not fight wars very well, or whether, instead, weak insurgencies are those incapable of enforcing hierarchy and therefore organize themselves as networks because they have no alternative. Insurgent philosopher/

---

[4]   Describing insurgent organization, Kulikov (2003, p. 22) states that

> Detachments generally are regional. Residents of a single village band together in so-called "self-defense detachments." Residents of a single county area are formed into "national militia brigades and regiments of Chechnya." These detachments fight only in the areas from which they are drawn.

[5]   Both Mao Tse-tung and Che Guevara describe a progression of force structure, although they envision development and end state occurring through somewhat different mechanisms. Mao also allows for dissonance in time and space: Portions of the insurgency may shift back and forth between phases in various places at various times in response to government activity and available support.

[6]   To winnow down these ambiguities to reach a binary answer for this portion of the study (hierarchical or networked), the survey asked the researchers whether the C2 for the *primary* insurgent group was "tight and hierarchical" or "more networked." As in all other sections of this study, the researchers investigated each case and made a subjective judgment based on the *preponderance* of evidence. An insurgency was judged to be hierarchical if the principal organization fit that description most of the time during most of the campaign. We judged the Chechen insurgency, with multiple leaders and units spread out over varying terrain and across at least two borders, to be networked. Cases that could not logically be divided were coded as "mixed."

practitioner Régis Debray (1967, pp. 72–73), a contemporary of Che Guevara, emphasized the necessity of a single insurgent command:

> The lack of a single command puts the revolutionary forces in a situation of an artillery gunner who has not been told which direction to fire. . . . The absence of a centralized executive (political and military) leadership leads to such waste, such useless slaughter.

In many cases, a hierarchical structure was necessary for the progression of an insurgency to victory. This phenomenon seems to point to the efficacy of hierarchical and (as we will show) rural insurgencies. We observed some qualitative trends across selected cases:

- Strength of the hierarchical structure did not offset the absence of reliable sanctuary and external sponsorship, two critical elements to insurgent success (e.g., the Malayan, Philippine Huk, and Sri Lankan LTTE insurgencies).[7]
- Some, but not all, successful networked insurgencies required at least a brief period of hierarchical organization to achieve victory. Therefore, most "pure" networked insurgencies never achieve absolute success (with Algerian independence perhaps serving as an exception). Even this temporary transition exposes the insurgency to at least a period of vulnerability. Galula points out that the Greek insurgency was defeated only after it organized along hierarchical lines and "accepted battle." He also believes that, "if the insurgent has understood his strategic problems well, revolutionary war never reverts to conventional form" (Galula, 1964

---

[7] Abdulkader H. Sinno (2008, p. 45) states,

> A safe haven is not essential to win the conflict; what is essential for the organization is to organize properly based on whether it has such a space. . . . An organization that suddenly gains control of a safe haven must centralize its structure to be able to take the strategic initiative. . . .

Sinno (2008, p. 41) also describes the resiliency of council-led insurgencies like the Quetta Taliban and AQI; this aspect of insurgent structure is worthy of additional study.

[2006], pp. 12, 15).[8] This places him at odds with Mao and most of Mao's disciples.

- In line with contemporary thinking, hierarchies are somewhat vulnerable to decapitation strikes. As we show later, hierarchical insurgencies led by highly charismatic leaders are most susceptible to decapitation. The Philippine Moro National Liberation Front (MNLF), Angolan UNITA, Turkish Partiya Karkerên Kurdistan (PKK, or Kurdistan Workers' Party), Sri Lankan LTTE, and the Peruvian Shining Path suffered either major setbacks or defeat when their leaders were captured or killed.[9] We also note that these defeats did not last if the government failed to address the root causes of the conflict.

- While hierarchies may be vulnerable to decapitation, networks do not afford watertight security for insurgency leadership or cadres. Uruguay's security services picked apart and essentially crushed the sizable Tupamaros organization in a relatively brief campaign, imprisoning most of its leadership. The Constitutionalist insurgency in the Dominican Republic was similarly defeated with U.S. assistance.

- Other insurgencies—especially those led by councils—prove capable of replacing leadership at various levels (not just the commander) without losing significant momentum. Palestinian Harakat al-Muqāwamat al-Islāmiyyah (Hamas, or Islamic Resistance Movement), Hizballah, Fatah al-Islam, AQI, and the African National Congress (ANC) in South Africa all managed to survive the loss or detainment of key leadership.[10] Marc Chernick (2007, p. 69) posits that the leadership council of the FARC could easily survive the loss of a single key leader.

---

[8] The loss of sanctuary in Yugoslavia and external support to the Greek government also strongly correlate with their defeat.

[9] Local culture often plays a role in the organizational structure of insurgent groups. Highly ordered and authoritarian societies often (but not always) produce hierarchical insurgencies. O'Neill (1990) discusses the effect of culture on organization.

[10] Sinno (2008, p. 78) states that "A governance board increases the resilience of a decentralized structure by making [it] even less useful to capture or assassinate its leadership."

- Networked insurgencies, especially those linked to urban civil uprising, seem to have less success achieving a victorious end state over a protracted campaign and more success in achieving quick victories. Although the Dominican Constitutionalists were eventually defeated, they did manage to overthrow the government early in the conflict, at the height of civil unrest.[11]
- Urban insurgencies tend to be networked, while rural insurgencies tend to be hierarchical. Not surprisingly, we found that the urban terrain necessitated the use of cellular structures.[12] In some cases (e.g., Nicaragua), the insurgency retained a hierarchical leadership in the countryside while relying on networked cohorts in the cities. In a few cases (e.g., Chechnya, at varying points along the conflict timeline), the reverse was also true. The impact of urban terrain on insurgent structures is explained in greater detail in the next section.

For operational purposes, definitively labeling an insurgent organization as hierarchical or networked in the absence of very clear and specific intelligence would be imprudent. In most of the 89 cases, the insurgencies possessed a core hierarchy supported by, or interlaced with, a diffuse network of militants, logisticians, financiers, and passive supporters. Some insurgencies survived the loss of the hierarchical core to subsist or, in some cases, to thrive as a network. In many cases, the insurgency sustained two distinct but interconnected organizations, epitomizing O'Neill's parallel-hierarchy structure. Parallel hierarchy can further stymie "line and block" analysis.

---

[11] Civil unrest is not a foolproof vehicle for the networked insurgent. It did not work in the Tibet case (1959).

[12] Various urban-insurgency tacticians both recognize this necessity and believe networks to be generally more effective than hierarchies. In a rather subjective treatise on urban insurgencies, James Kohl and John Litt (1974) document a range of these viewpoints and arguments. Also see Marighella (1969 [2008]).

Thailand and the Barisan Revolusi Nasional and the Patani United Liberation Organization

Thailand's insurgency provides interesting anecdotal evidence of these complexities. Two separate Malay Muslim insurgent groups fought against the Thai government from the 1960s through the late 1980s. Both groups, the Barisan Revolusi Nasional (BRN, or National Revolutionary Front) and the Patani United Liberation Organization (PULO) relied on hierarchical lines of command. Both groups achieved some initial success, but, by the mid-1980s, the insurgency was faltering. The two groups decided to form a loose coalition in order to revitalize their respective movements. "New PULO," a splinter organization, joined them. These three groups essentially formed an insurgent network while retaining their original, internal hierarchical structures (Chalk, 2008, pp. 1–10).[13]

The Thai government responded to this new alignment by convincing neighboring Malaysia to deny sanctuary to the insurgents. This strategy had immediate effect, shrinking insurgent supplies and sapping morale. Within a few years the BRN, PULO, and New PULO ceased to exist as legitimate military organizations or political movements. However, the Thai government failed to capitalize on this victory with either economic-development packages or political overtures to the dispossessed south. A police crackdown ensued. In reaction to this anocratic behavior, the Malay Muslim insurgency reignited circa 2001. Neither the BRN nor the PULO movements resurfaced in strength, however (see Figure 4.2).[14] Instead, the insurgency existed in early 2009 as a loosely confederated movement of independent cells consisting of BRN and PULO remnants, other small insurgent groups, and pockets of disaffected youths (Chalk, 2008, p. 9).

Until it coalesces around a centralized hierarchy, this insurgency is likely to achieve little more than sustained chaos in southern Thai-

---

[13] Sinno (2008, p. 36) states that these kinds of "superorganizational institutions," while survivable, are relatively ineffective.

[14] PULO continues to issue statements and take credit for attacks, not all of which are verified.

Figure 4.2
False Endings of Muslim Malay Insurgency

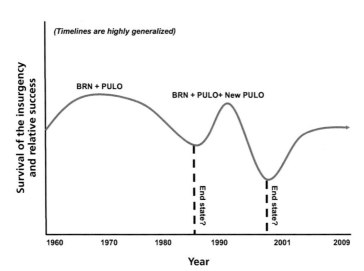

land.[15] Conversely, the Thai government would be hard pressed to dismantle the elusive and perhaps nonexistent insurgent "military wing" by tactical means. In fact, efforts to do so have resulted in collateral damage that, in turn, undermined support for the central government. Through 2007,

> the hard military track was not accompanied by a softer, more nuanced policy line to win popular support. The Thai government made virtually no effort to address the poverty, underdevelopment, and general alienation that fuels Malay-Muslim discontent. (Chalk, 2008, p. 18)

Thirty-nine years after its inception, the Thai insurgency appears to have no end in sight.

That the Thai government ever had an opportunity to end the insurgency with a decapitation strike or through the use of effects-

---

[15] Between 2004 and 2009, violence increased but was variously attributed to different groups and subgroups.

based tactics is debatable. At no point along the conflict timeline would the death of a single leader, or even a group of leaders, have salved the disenfranchisement of the Malay Muslim population. In the first stage of the insurgency, from 1960 through the mid-1980s, the destruction of any of the three separate groups likely would have strengthened the remaining groups and might have sent surviving leadership deeper underground. Simultaneous and complete destruction of all three groups would probably have succeeded in sending the insurgency into remission, but little more.

In the second stage of the insurgency, the coalition was, in essence, leaderless. It is unlikely that the Thai government could have shattered this leaderless coalition through targeted killings of key communicators, and, if it had, the three groups would have continued to operate independently.[16] In its current incarnation, the insurgency appears to have morphed into a disassociated manifestation of popular discontent. Military action in this case is little more than repressive population control.[17]

In other cases, we found that insurgent leaders may wish to give the impression of strict hierarchy where it does not necessarily exist: Captured insurgent line-and-block charts may represent ghost cadres.[18] In some cases, insurgent leaders may be under the false impression that their organization is hierarchical down to the smallest cell, when, in fact, their ability to command and control is restricted by terrain, communications, personality, or even internal corruption. Lower-level leaders often have personal or local agendas tied to tribal obligations, criminal activities, or internecine antagonisms with insurgent peers. These local leaders may accept formal orders and submit formal reports

---

[16] In another case of fractionated insurgent movements, the India-Naxalite insurgency may have consisted of 30 active groups at one point, but certainly consists of at least ten (Judge, 1992, p. 44).

[17] Chalk (2008, pp. 13–22) discusses the possibility that the Thai insurgency is transforming into a proxy for international Islamic terrorism. He believes that the conflict will retain its regional focus. He also believes that the insurgency is reaching a tipping point and that it may "spiral out of control." Kilcullen (2009, p. 213) paints the Thai-Malay insurgents as "accidental guerrillas" rather than radical Islamists.

[18] Author has direct knowledge of such cases in Iraq, circa 2006.

while quietly conducting independent operations that do not match the intent of the central leadership.

Some generalizations can be drawn from this analysis: First, it is exceedingly tricky to "map" an insurgent organization with accuracy—any intelligence line-and-block chart should be presented and viewed with a healthy degree of skepticism. Second, organizations fluctuate in both form and function over the course of a conflict; third, it is very difficult (but not impossible) to force an ending by decapitation. In terms of bringing a conflict to an end, it is clear that a decapitation strike of even hierarchical organizations does not guarantee success.[19] As we have identified, most insurgencies are resilient, either suffering long death throes or making an eventual comeback. Continuing government repression in the absence of total victory allows the embers of the insurgency to smolder, perhaps indefinitely.

Further, targeted raids conducted in the absence of dedicated, effective reforms (or significant happenstance) also lack efficacy in many, but not all, cases. Insurgencies—especially rural insurgencies—are typically more complex and more extensive than isolated terrorist movements in that they often represent broad grassroots sentiment. Tactics that prove successful in defeating finite terrorist networks (e.g., targeted raids) seem in general to be less effective against insurgency movements. As proven in Nigeria, Sri Lanka, Vietnam, Iraq (2003–), and Afghanistan (2001–), targeted raids that involve the use of indirect fires (air, artillery, missiles) or home invasions also carry the risk of alienating both the local population and, in the case of intervention, domestic political support. In each of these cases, the reaction against these activities undermined civil and political lines of operation.[20]

---

[19] We present data on decapitation strikes later in this monograph.

[20] This proved true across a range of other cases as well. Often, accounts of civilian casualties published in the wake of aerial bombing attacks proved inaccurate (see, for example, Connable, 2009). Even inaccurate stories undermined the counterinsurgent. Drone attacks in Pakistan have led even some leading U.S. COIN experts, including Kilcullen, to question aerial targeting tactics (Kilcullen and Exum, 2009).

Urbanization

Our study shows that, in countries with less than 40 percent urbanization, the government lost about 75 percent of the time (8:19), with other cases ongoing or mixed (see Table 4.2). Its win ratio rose to almost 3:1 when urbanization was between 40 and 70 percent (14:5) or higher than 70 percent (6:2), also stripping away ongoing or mixed cases. Insurgencies appear to be far more successful in the countryside than in cities (see Figure 4.3).

In the mid- to late 20th century, many insurgencies developed around issues of land reform, a populist (and popular) theme.[21] A range of communist insurgent philosophers, including Mao Tse-tung and Che Guevara, harped on the adoption of land reform as a fundamental insurgent goal. They did so with good reason: Countries heavily populated by an impoverished rural peasantry provide ripe

Figure 4.3
Urbanization

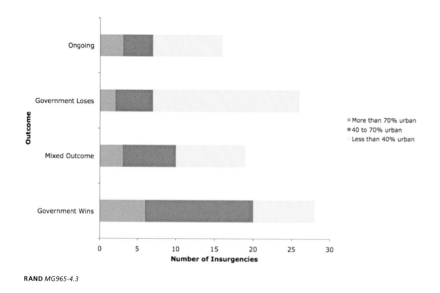

RAND MG965-4.3

---

[21] Even if the specific theme of land reform is not as common in insurgent propaganda today as it was in the 1960s, the general theme is still valid and widely used.

Table 4.2
Number of Insurgencies by Level of Urbanization

Outcome	More Than 70% Urban	40–70% Urban	Less Than 40% Urban
Government wins	6	14	8
Mixed outcome	3	7	9
Government loses	2	5	19
Ongoing	3	4	9

ground for insurgents. Even if the specific theme of land reform is not as common in insurgent propaganda today as it was in the 1960s, the general themes of rural justice and equality are still valid and widely used. While the Frente Sandinista de Liberación Nacional (FSLN, or Sandinista National Liberation Front) of Nicaragua built both rural and urban constituencies, at various points during the conflict, it preached land reform to gain support from the rural poor.[22] The Unified Communist Party of Nepal incorporated land reform into its platform, referring to government practices as "feudal."[23] Charu Mazumdar and Kanu Sanyal, Indian communist ideologues and leaders of the Naxalbari insurgency, also drew on feudalism as a fundamental propaganda theme.[24] Shortly after taking power in Cuba, Fidel Castro issued a sweeping agrarian reform law that redistributed large, private

---

[22] The FSLN overthrew the Nicaraguan government in 1979. Kohl and Litt (1974, p. 26) hypothesize that this kind of "combined struggle," consisting of both rural and urban operations, run by a hybrid networked-hierarchical leadership, is necessary for the insurgent to achieve victory: "Combined struggle is the synthesis unifying the contradictions between city and countryside . . . and a vital step towards achieving 'the people in arms.'" Some expert discussion revolves around a hybrid *counterinsurgency* model designed to address a full-spectrum conflict like Iraq.

[23] The Maoist 40-point memorandum to Nepal's prime minister was published on February 4, 1996. Communist revolutionaries make liberal use of the word "feudal" in speeches and in operational dicta.

[24] The Naxalbari movement erupted in Western Bengal, India, in the late 1960s. It is commonly referred to by the descriptive *Naxalite*. Both the Nepalese and Naxalite Communists are considered Maoist, although Naxalite philosophy identified *foquismo* as a central operational tenet.

land holdings to the masses. The Colombian FARC insurgency is portrayed as a struggle between landed gentry and rural agrarian peasants (Chernick, 2007, p. 60). General Võ Nguyên Giáp, mastermind of the successful Viet Minh campaign against the French and, later, the Vietcong and NVA campaigns against the Americans and South Vietnamese, relied on land reform as a central propaganda theme (also using a translation for the term *feudal*) in *People's War, People's Army* (Võ, 1961 [2000]).[25] Che Guevara (1969 [2008], pp. 12–13) sums up their conventional wisdom:

> [T]he guerrilla fighter will carry out his action in wild places of small population. Since in these places the struggle of the people for reforms is aimed primarily and almost exclusively at changing the social form of land ownership, the guerrilla fighter is above all an agrarian revolutionary.

Urban insurgencies or those principally focused in urban centers are, by definition, removed from the rural peasantry. Land reform and other populist rural themes may figure prominently in the urban insurgent's propaganda campaign, but they probably do not resonate with the same depth as in the countryside. O'Neill (1990, p. 45) explains why some insurgents choose to operate in urban terrain. In some places,

> inaccessible rural areas where guerrillas may operate with impunity simply do not exist. Accordingly, insurgents who pursue political aims through violent acts have been compelled to locate in the cities and to operate on a small scale in order to survive.

Urban terrain is naturally tight, and urban populations are able to communicate and interact with rapidity unavailable to the typical rural insurgent. Kohl and Litt (1974) describe and contextualize various late-1960s theories of urban insurgency. They attribute the near total failure of Guevara (post-Cuba) and others to rural insurgents' vulnerability to contemporary COIN tactics and technologies:

---

[25] Giáp is the author's given name in Vietnamese.

> Stalemated in the isolated and often inhospitable countryside, guerrilla warfare was reconceived in an urban milieu. Here again the advantages of surprise and initiative, clandestinity [sic] and mobility, could be reasserted. . . . The urban guerrilla also enjoyed an initial advantage owing to the fact that counterinsurgency techniques evolved to meet the revolutionary challenge in the countryside were largely inapplicable in the city. (Kohl and Litt, 1974, p. 10)[26]

Urban terrain confers additional benefits to insurgents. Interconnectedness and physical immediacy gives them the ability to quickly amass the sizable concentrations of people necessary to support palace coups or to foment demonstrations and other civil unrest. As the British COIN manual states, "The ready availability of large numbers of people means that a crowd can be assembled and demonstrations engineered, with comparative ease; these can then be manipulated" (UK Ministry of Defence, 2001, p. A-3-11).

This was proven in the first stage of the Dominican case, in the Tibetan case, and in some of the final stages of other insurgencies. Insurgents and supporters can just as quickly disperse in order to avoid immediate retribution. Guillén (1973, p. 240) describes urban insurgents who "[l]ive separately and fight together in order to elude police repression." Marighella (1969 [2008], p. 23) describes the life of the urban insurgent: "The urban guerrilla must know how to live among the people, and he must not appear strange and different from ordinary city life."

Static security-force positions provide inviting targets to the urban insurgent. During the Algerian insurgency against the French (1954–1962), the insurgents capitalized on the vulnerability of security forces in the urban canyons by conducting a merciless and effective series of terrorist attacks and raids against static positions. The Chechens, experts in urban ambush, assassination, bombing, and kidnap-

---

[26] Here, Kohl and Litt seemed to parrot Guillén (1973, p. 234), who makes a nearly identical argument. Kohl and Litt's analysis did not necessarily account for advances in modern technology (e.g., unmanned aerial vehicles and advanced communication intercepts), but urban terrain continues to offer a greater degree of anonymity than rural terrain.

ping techniques, scored a number of dramatic successes against Russian security forces in both Chechnya I and II.

On balance, however, the counterinsurgent seems to draw greater advantage from urban terrain. We found that, in many but not all cases, the longer an urban insurgency lasted, the easier it became for the government to zero in on insurgent leadership and cadres.[27] Tight quarters and rapid communication provide security-service informers ample opportunity to observe and track insurgent activities. In describing the faltering early stages of the Sandinista campaign in Nicaragua, Hammes (2006, p. 77) states,

> First, they flirted unsuccessfully with an urban revolution. This, too, failed—in the same way it failed in other nations. The entire theory of forming guerrilla forces in the cities played to the strengths of the dictator's security forces.

Capable police forces proved critical in each case in which counterinsurgents defeated an urban insurgency. Police were especially effective in urban environments at ferreting out insurgents and in penetrating insurgent networks.[28] Marighella (1969 [2008], p. 36) admits as much, stating, "In this (urban) conflict, the police have superiority." There are few moments of respite for the urban insurgent. When insurgents were identified, reaction forces quickly isolated and attacked their safe houses.[29] Rarely can insurgents rely on consistent, voluntarily

---

[27] Governments that failed in urban terrain (e.g., Algeria 1954–1962) typically undermined their own tactical advantages by employing excessively violent COIN techniques. Qualitative analysis shows that, in some cases, however, these techniques are successful: Violent repression proved to be a winning tactic for the Uruguayan government against the Tupamaros.

[28] In an urban insurgency, "[t]he government's [principal] sources of information are infiltrators, defectors [and] informers" (Kohl and Litt, 1974, p. 21). Galula (1964 [2006]), Gurr (1974), Kilcullen (2009), Clutterbuck (1966, 1977), and O'Neill (1990) also emphasize the value of police and police intelligence.

[29] Safe houses should not be equated with safe havens. Safe houses are tactical and disposable, while safe havens (sanctuaries) are strategic and strategically vital. FM 3-24 describes "virtual" sanctuaries and states that urban terrain provides safe haven. These notions are too esoteric to be applied to the historical cases in our sample, and we do not concur that urban

provided haven in urban or even suburban terrain. Lack of haven either forces them into the rural areas or leaves them vulnerable to infiltration and raids. As we identified in the previous section, this leads many urban insurgencies to rely on complex, and often unwieldy, networked C2 relationships.

Some conventional wisdom suggests that, even when the urban insurgent can overcome security-force restrictions, the inherent flaws in the urban-pure model continue to generate failed end states. The British COIN manual comments,

> (The) fatal flaw in any urban guerrilla strategy is that it lacks completeness. The theory is that when the guerrillas have succeeded in driving the government into a sufficiently repressive posture, the populace will rise up in righteous wrath and destroy its oppressors. But even if the population should decide that it is the government and not the guerrillas that is responsible for its growing misery, there is no plan of how to eliminate the government. (UK Ministry of Defence, 2001, p. A-1-A-6)

Intrinsic to this "fatal flaw" is the seeming inability of the urban insurgent to build genuine popular support, perhaps the most critical element of success for any insurgency. In each case of urban insurgency identified in this study, propagandists faced exponentially more-complex audiences in cities than they would typically face in the country. City dwellers lack the common identity and motivations found in tribal villages. This also holds true for urban slums, but to a lesser extent. An insurgency's message and base of support is therefore likely to be more diluted among an urban population, weakening the group's power and resilience. This last point is probably the most pertinent when considering the impact of terrain on insurgency endings. Again, the Tupamaros case is relevant.

---

terrain equates to sanctuary in modern COIN. It protects some insurgents from some methods of attack (i.e., aerial attack, sometimes) but not others (i.e., police intelligence operations). Guillén (1973, pp. 264–265) believed that safe houses tie insurgents to fixed terrain, eroding mobility and security. He stated, "Houses that serve as barracks or hideouts tend to immobilize the guerrillas and to expose them to the possibility of encirclement and annihilation."

Uruguay and the Tupamaros

The Tupamaros began planning and fomenting their insurgency in 1963, riding a revolutionary "wave" created by Fidel Castro's success in Cuba in 1959.[30] Their goal was to replace the Uruguayan government, a regime they viewed as increasingly anocratic and corrupt, with a rather fanciful communist system of self-management. In some respects, the Tupamaros followed the Guevarist *foco* model of revolution, in that they believed that a small, popular insurgency would breed mass uprising.[31] Tupamaros adviser Abraham Guillén envisioned an urban *foco*.[32]

Guillén saw the possibilities of (as O'Neill, 1990, p. 52, describes it) a "Leninist type of takeover."[33] Uruguay's geography was not suited to a rural insurgency in any case. Wright (1991, p. 97) states that Uruguay had "the least hospitable terrain in all of Latin America for rural *focos*." Literally half of the country's population lived in the capital, Montevideo (O'Neill, 1990, p. 52; Wright, 1991, p. 97).[34] The Tupamaros focused nearly all their attention on the capital, with some vague commitments to support rural movements and with an equally vague intention of using the urban insurgency to spark a more general rural uprising (O'Neill, 1990, p. 46).[35]

---

[30] The group's official name in Spanish was *Movimento de Liberacion National*, or "national liberation movement." They were popularly referred to as the Tupamaros, a probable contraction of "Tupac Amaro," an Incan notable murdered by the Spanish in the 16th century.

[31] Kohl and Litt (1974, pp. 229–230) translated an interview with a Tupamaros leader initially published in a Chilean journal in 1968. The Tupamaros interviewee clearly enunciates belief in *foco* theory.

[32] Abraham Guillén might be considered the Tupamaros' grand strategist. He was an educator, reporter, and intellectual.

[33] Carlos Marighella led an insurgency in Brazil, where he was killed in 1969. Marighella wrote the *Minimanual of the Urban Guerrilla* (1969 [2008]) and *For the Liberation of Brazil* (1971).

[34] This reality was also reflected in Kohl and Litt (1974) across each case study, but particularly in the Uruguay case.

[35] O'Neill interprets Guillén and Marighella as viewing urban insurgency as a phase; the insurgency would eventually move into the countryside in latter stages. Based on our research, we find it possible that Marighella simply paid lip service to the idea of rural action, while Guillén formulated his theory of hybrid, urban-centric insurgency somewhat in retrospect. While Guillén described a hybrid campaign of mutually supporting rural and urban

It is difficult to trace the Tupamaros insurgency in the generally clear, curving lines to end state as we have with some other cases (see Figure 4.4). The Tupamaros' fortunes "fluctuated between spectacular successes and severe setbacks" (O'Neill, 1990, p. 100).[36] Events along the conflict timeline were smaller in scale and much shorter lived, but the urban terrain magnified the effect of each event or series of events on the rise and fall of the insurgency. As with each previous case, the timeline reveals only a small set of causative factors that led to the end of the insurgency. The Tupamaros alienated the populace with a series of extreme efforts to ignite the urban *foco*—in a few cases, forcing civilians to endure propaganda speeches at gunpoint. Their unwillingness to negotiate over hostages led some members of the public to hold the Tupamaros responsible for the oppressive military crackdown. Their

Figure 4.4
Rise and Fall of Urban Tupamaros

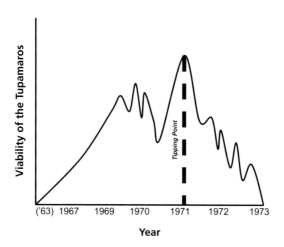

NOTE: Plots are exaggerated to show the effective increase in frequency and amplitude of significant shifts in momentum.
RAND *MG965-4.4*

---

operations, he seemed to press hard for, and to see greater value in, urban action. In a characteristic volte-face, Guillén (1973, pp. 256–257) takes Marighella to task for supporting (we believe dishonestly) a hybrid urban-rural theory while defending urban operations.

[36] The timeline is drawn from O'Neill (1990, pp. 100–101).

excessive focus on violent action probably contributed to the general alienation of their target audience. The United States provided some indirect support to the government, greatly strengthening the hand of the security services. Finally, brutal security-service tactics proved effective against the often-unpopular urban insurgents.

Lasting ten years, the Tupamaros insurgency appears on the surface to be an "average" case study.[37] A thorough examination of the insurgency's timeline and end state, however, reveals some significant variations. We have already identified the high-frequency, high-amplitude sine wave of events over the ten-year timeline as divergent from the more common rural insurgencies.[38] By comparing this timeline to other urban insurgencies and insurgencies with large urban components, a trend seems to appear. High-tempo operations in tight terrain offer a few opportunities for rapid success and the omnipotent specter of precipitous failure. For a brief time, the Tupamaros represented a serious threat to the regime, but they fell quickly once the tipping point had been reached. This "quick tip" bucks the trend of the long end-state tail we identified in previous sections.[39] Guillén (1973, p. 274), a theoretician and rather sharp critic of insurgent strategies, bluntly summarizes the Tupamaros insurgency as follows:

> The Tupamaros were the first group of urban guerrillas to teach the world how to initiate an insurrection in the cities with few supporters and modest means. But their superb tactics have been nullified by a mediocre strategy and a questionable politics.

As morally repugnant as they may be, extremely violent repressive tactics also appear to be more effective for the counterinsurgent in urban terrain. These tactics proved essential in quelling, driving underground, or completely crushing insurgent cadre in Chechnya (II),

---

[37] Ten years is our statistical median, not statistical mean or average. "Average" is a qualitative modifier in this sentence.

[38] A detailed 30-page timeline is available in Kohl and Litt (1974, pp. 196–226). If graphed, the detailed timeline would appear saw-toothed.

[39] Although we did not explore the Brazilian insurgency, the timeline and outcome dynamics of Marighella's movement effectively parallel those of the Tupamaros.

Argentina, Uruguay, Tibet, Algeria (Groupe Islamique Armé, GIA, or Armed Islamic Group), and Peru.[40] This may prove true for a variety of reasons. First, as we discussed, insurgent organizations are more vulnerable in urban areas. Second, all counterinsurgent tactics are generally more effective when the insurgent is not availed of safe haven (4:1 advantage for the counterinsurgent, with other cases ongoing or mixed). Third, urban insurgents seem to rely more heavily on violent action and terrorism than their rural counterparts, a pattern that may give credibility to, or even excuse, some government violence.[41]

Uninhibited violence stands a fair chance of quickly crushing an "urban-pure" insurgent military organization. This trend did not prove true in cases in which the insurgency could rely equally on urban and rural support (e.g., Guatemala 1960–1996, Nicaragua Sandinistas). In those "hybrid" cases, when insurgents suffered a defeat in the urban areas, they could fall back into their safe havens to recover or hibernate.[42] The absence of rural support leaves the urban insurgent vulnerable to total elimination. The authors of the British COIN manual make an interesting observation about the Cuban case in a discussion on *foco* theory, one that does not surface in many more-superficial treatments of the subject:

---

[40] In the section on anocracy, we describe how these violent tactics may succeed in establishing short-term stability but rarely address the root causes of the insurgency. Extreme repression on either side usually backfires in either the short or long run.

[41] While Mao, Che, Giáp, and Guillén preached population-centric warfare, in the *Minimanual*, Marighella (1969 [2008], p. 24) stated that violent action was the central tenet of an urban insurgency. He identifies his "two essential objectives" as (1) the physical elimination of the leaders and assistants of the armed forces and of the police and (2) the expropriation of government resources and the wealth belonging to the rich businesspeople, the large landowners, and the imperialists, with small expropriations used for the sustenance of the individual guerrillas and large ones for the maintenance of the revolutionary organization itself.

[42] T. X. Hammes and David Kilcullen both address hybrid insurgencies and COIN strategies. In cadence with Guillén, they describe hybrid insurgency as targeting both rural and urban populations with relatively equal focus. In western COIN circles, the term *hybrid* is increasingly used to describe collateral application of asymmetric and conventional (or multitiered asymmetric) tactics and strategies (e.g., guerrilla warfare and combined arms warfare). See McCuen (2008, p. 107) and F. Hoffman (2009).

The emphasis Guevara placed upon rural operations grossly underestimated the extent to which Castro's victory had actually depended upon the contribution made by urban groups; the latter not only supplied the Rebel Army with recruits and arms but also prevented Batista from devoting his full resources to the campaign against the Sierra-Maestra based insurgents. (UK Ministry of Defence, 2001, p. A-1-D-1)

The kind of hybrid urban-rural strategy promulgated by Guillén and practiced in Nicaragua seems to hold much greater promise for insurgents than either urban- or rural-pure insurgencies. Although somewhat prone to rhetorical digression, Guillén (1973, p. 244) made several simple yet salient points about hybrid warfare and contextualized them in contemporary experience:

The secret of revolutionary victory lies in the unity of country and city under the same strategic direction in the revolutionary war. . . . [T]he guerrillas have to change their tactics according to the terrain: in open countryside they should work by day and fight by night; in the forests and mountains the struggle is a continuing one, with the possibility of establishing liberated zones in dense forests at high altitudes; in the cities the guerrillas agitate, fight and give cover to the masses, but cannot establish liberated zones until there is no longer danger that the enemy will surround, bombard, and annihilate them.

Finally, O'Neill (1990, pp. 57, 160) provides a handy segue into our discussion on insurgent use of terror:

[While] cities may provide opportunities for terrorists whose operations are ancillary to rural warfare, they have not, by themselves, proven to be areas where decisive strategic successes can be achieved against committed governments with adequate resources. . . . [I]nsurgents who follow an urban warfare strategy and emphasize terrorism rather than organizing popular support may be countered by a modest but vigorous program centered on intelligence, police, and legal due process.

Insurgent Use of Terrorism

Today's insurgencies often combine traditional actions and the kind of terrorism with which the 21st century may come to be associated. But does terrorism do insurgents much good? We found that the reverse is more likely to be true (see Figure 4.5). It is also worth noting that insurgencies that employ terror broadly are more likely to end quickly.

On the surface, insurgent dogma on the use of terror is characteristically Manichaean, with only a few groups equivocating on the subject. When insurgents publicly condoned terrorism, they unfailingly followed through with action. Other groups denounced terror in public but would often excuse or make use of generally recognized terror tactics over the course of the conflict. Justifications for terror vary from group to group (and definitions from historian to historian), muddling efforts to explain insurgent behavior. Some insurgent groups took a middle ground, permitting terrorist attacks against certain groups but not others. Conventional wisdom divides on the subject of terror, and a review of a few dichotomous insurgent philosophies is

Figure 4.5
Insurgent Use of Terrorism

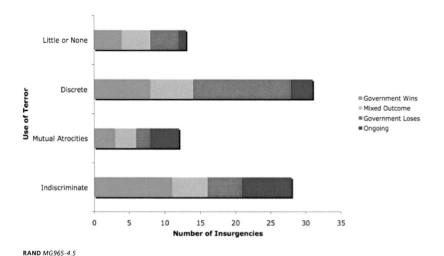

RAND MG965-4.5

helpful to set the stage for case-study discussion.[43] Writing in the wake of the successful Cuban campaign, Che Guevara (1969 [2008], p. 100) expressed the opinion that terrorism was not only unnecessary but also counterproductive:

> [T]errorism and personal assaults are entirely different tactics [from sabotage]. We sincerely believe that terrorism is of negative value, that it by no means produces the desired effects, that it can turn a people against a revolutionary movement, and that it can bring a loss of lives to its agents out of proportion to what it produces.

Disagreement over tactics reveals further division between Guevara and Marighella. In keeping with his focus on tactical violence, Marighella (1969 [2008], pp. 20, 66) viewed terrorism not only as justifiable but also as "ennobling":

> The accusation of "violence" or "terrorism" no longer has the negative meaning it used to have. It has acquired new clothing, new color. It does not divide, it does not discredit; on the contrary, it represents a center of attraction. Today, to be "violent" or a "terrorist" is a quality that ennobles any honorable person, because it is an act worthy of a revolutionary engaged in armed struggle against the shameful military dictatorship and its atrocities. . . . Terrorism is a weapon the revolutionary can never relinquish.

Marighella winks at the use of torture to obtain confessions from captured security-service personnel. Abimael Guzmán, leader of Peru's Shining Path insurgency from 1980 until his arrest in 1992, expressed similar sentiment:

> We start from the position that we do not subscribe either to the Universal Declaration of Human Rights or the Costa Rica Dec-

---

[43] Mao Tse-tung generally avoids the subject in *On Guerrilla Warfare*, but, in some of his other writings, he denounces terror against civilians as counterproductive. In practice, he brutalized even his own insurgent armies while fighting the Japanese occupying army and the Kuomintang (China 1934–1950) and, later, during his chairmanship of the PRC.

laration [American Convention on Human Rights] . . . Shining Path's position is quite clear, we reject and condemn human rights because they are reactionary, counter-revolutionary, bourgeois rights; they are presently the weapon of revisionists and imperialists, principally of Yankee Imperialism.[44]

Guzmán's organization was notorious for committing atrocities, many of which were documented in the Amnesty International report. Hareth al-Dhari, leader of the Association of Muslim Scholars (AMS, transliterated from Arabic as *Hayat al-Ulema al-Muslimin*) and a guiding light for various Iraqi nationalist insurgent movements (Iraq 2003–), publicly denounced the use of indiscriminate terror against civilians:

> Those who target innocent and peaceful Iraqis from all sects, denominations, and faiths are condemned criminals that transgress against Islamic Jurisprudence (Shari'ah) [sic] and are outside the law and the national values. They are like the enemies and occupiers of the homeland regardless to which sect or faction or faith they belong.[45]

Keeping in mind that no one factor can be identified as causative, the relative success and failure of each of these philosopher/practitioners may show some qualitative correlation between the use of terror and end state:

- Che Guevara saw mixed results. His philosophy was proven in Cuba and then disproven in the disastrous *foco* campaigns that followed.

---

[44] In its review of the human-rights abuses committed by both sides during the Shining Path insurgency, Amnesty International cites Guzmán's "Above the Two Hills: Counterinsurgency War and Its Allies," published in 1991.

[45] The Association of Muslim Scholars website carries an article quoting al-Dhari (2007) titled "To Attack Innocent Civilians is Not Jihad!" Insufficient unclassified documentation exists to prove that Al-Dhari privately supported the use of terrorism or torture in the later stages of the conflict.

- Marighella's violent urban insurgency was defeated, and he was shot by Brazilian police in Sao Paolo in 1969.
- Peruvian security services captured Guzmán in 1992, sending the Shining Path insurgents into hibernation for a decade.
- One could argue that Hareth al-Dhari succeeded. By early 2009, thousands of nationalist insurgents had been inducted into the local police and militias, and the Americans had pledged a firm date of withdrawal.

Of the four, only al-Dhari survived the conflict both physically and politically.

Some counterinsurgents concur with Che. Galula (1964 [2006], p. 60) sees short-term efficacy and long-term risk for the terrorist-insurgent:

> By terrorism small groups of insurgents have been catapulted overnight to the top of large revolutionary movements, and some of won their victory at that very time, without need for further action. However, the bill is paid at the end with the bitterness bred by terrorism and with the usual post-victory disintegration of a party hastily thrown together.[46]

O'Neill also states that terror tactics can help insurgents progress their agendas. His only caveat, however, is that the agendas may be multifaceted and opaque to the counterinsurgent:

> Terrorism is a form of warfare in which violence is directed primarily against non-combatants. . . . Insurgent terrorism is purposeful, rather than mindless, violence because terrorists seek to achieve specific long-term, intermediate, and short-term goals. (O'Neill, 1990, p. 24)

McCuen treats the subject with some depth, describing terrorism as a tool of the weak. He identifies terror as a phase of insurgency,

---

[46] He also warns against government use of terror: "Can the government use terrorism too? It would be self-defeating since it is a source of disorder, which is precisely what the counterinsurgent aims to stop" (Galula, 1964 [2006], p. 74).

placing it before guerrilla warfare. Here, he quotes a 1951 directive issued by the Malayan communists who were, at the time, struggling to regain the initiative from the British. The insurgents perhaps realized that their last-ditch terror tactics were alienating the population:

> Party members are reminded that their primary duty is to expand and consolidate the organization of the masses, which is to take precedence over the purely military objective of destroying the enemy. . . . To win the masses the party must (i) stop seizing identity cards . . . ; (ii) stop burning new villages . . . ; (iii) stop attacking post offices, reservoirs, power stations, and other public services; (iv) refrain from derailing civilian trains with high explosives; (v) stop throwing grenades and take great care, when shooting [British sympathizers] found mixing with the masses, to prevent stray shots from hurting the masses; and (vi) stop burning religious buildings, sanitary trucks, Red Cross vehicles, and ambulances. (McCuen, 1966, p. 151, quoting a December 1, 1952, article from London's *Times*)[47]

How did various approaches to terror affect the tipping point, ending, and final outcome of the inclusive cases?[48] We found that the statistical differences between the indiscriminate and discrete use of terror were most telling. Marighella and others proposed the use of terrorism as a means to undermine the government and to weaken popular support for the ruling political elite. In many cases, however, indiscriminate terror had the inverse affect. Populations exposed to atrocities often split deeply along regional, religious, class, or ethnic lines, hardening their positions as the violence swelled. Indiscriminate tactics sometimes eroded the popular support for insurgencies among the populations most likely to back their cause. Abu Musab al-Zarqawi, the now-deceased leader of AQI, intentionally alienated Iraqi Shi'a but

---

[47] This directive is remarkably similar to a Quetta Taliban directive on the same subject.

[48] Kalyvas' (2006) *The Logic of Violence in Civil War* is essentially dedicated to the study of selected and indiscriminate violence, although he focuses on both insurgent and government violence. *The Logic of Violence in Civil War* contains one of the best available sets of data on the subject. Weinstein (2007) addresses the same subject but relies more heavily on qualitative analysis and case studies, also to good effect.

also lost the support of Iraqi Sunni through the use of a wide range of indiscriminate terror tactics.[49]

The factor of time adds some nuance to this assessment: The longer the insurgents employed indiscriminate terror, the more likely it seemed that the population would tip against the insurgency or in favor of the government.[50] This dynamic was sometimes offset when both parties employed indiscriminate terror tactics (e.g., Algeria 1954–1962), but not always. As we have shown in the Uruguay case and as we will show in the Peru case, the government's repressive activities proved highly effective against indiscriminately violent insurgencies.

We stated in the introduction to this section that insurgencies practicing indiscriminate terror tended to end quickly. This especially proved true in the absence of a stabilizing force (e.g., UN peacekeepers) or brokered peace processes. Lengthy campaigns of terror were likely to draw the attention of the international community or invite external intervention on behalf of the government. Neither of these eventualities tended to serve insurgent goals. When the international community became engaged in peace talks or negotiations, the insurgencies tended to tail out with mixed, inconclusive, ongoing, or marginally decisive endings. This proved true in Northern Ireland, Rwanda (1990–1994), Kosovo (1996–1999), Bosnia (1992–1995), and Sierra Leone (1991–2002).

Abimael Guzmán Reynoso's Shining Path insurgents fought to overthrow the government of Peru between 1981 and 1992.[51] An ardent communist and self-identified champion of the proletariat, Guzmán focused his propaganda against the urban elite.[52] The Peru-

---

[49] Criminal corruption and aggressive black marketeering helped undermine the AQI franchise in the eyes of Sunni Iraqis.

[50] This is a qualitative rather than quantitative finding, in that our statistical analysis did not address the factor of time in association with the coded data for indiscriminate terror.

[51] In Spanish, Shining Path is *Sendero Luminoso*. While we cover the period between 1981 and 1992, the Shining Path insurgency actually tailed out beyond 1992 and reerupted in the first decade of the 21st century.

[52] He included the urban slums in Lima as potential recruiting grounds and sources of popular support. This support never materialized for a variety of reasons, many cultural but some related to Shining Path tactics and hard-core propaganda.

vian capital of Lima, his ultimate objective, was a sprawling metropolis that was, in many ways, insulated from the rural population. Employing what could be described as a hybrid approach, Guzmán first rallied the countryside by targeting government and military officials in isolated towns and villages. He quickly expanded his operations to the capital, and, by 1983, he had conducted several spectacular attacks in and around Lima. With only a few statistical anomalies, the number, complexity, and audacity of the group's attacks increased steadily from 1981 through 1991. By 1991, the insurgents effectively controlled a wide swath of the countryside and had severely disrupted daily life and government functions in the capital (McCormick, 1992).

As Shining Path expanded its operations, it also accelerated the practice of torturing and killing civilians—local militia members as well as unarmed women and children. Sixty percent of recorded Shining Path attacks in Lima reportedly were carried out by bombings, a tactic that sometimes wounded or killed unfortunate pedestrians. While they primarily targeted security forces, government offices, and government agents in Lima (42 percent of all attacks), they also targeted civilian institutions and businesses. In the countryside, they aggressively targeted civilians: A full 45 percent of recorded attacks on the various rural fronts were conducted against "social" targets (civilians) or against civilian businesses (McCormick, 1992, pp. 35–36). A number of these attacks were well-publicized massacres.

Between 1981 and 1983, the government responded in kind as it struggled to catch up with an insurgency it had until then all but ignored. Peruvian security forces added a chapter to the long and sordid history of repressive COIN campaigns fought throughout Latin America between 1960 and the mid-1990s. Under emergency legal authority, military and paramilitary troops tortured, killed, and, in some instances, raped both suspected insurgents and civilians. By late 1983, however, the Peruvian military leadership tried to limit atrocities, implementing a traditional civic-based COIN strategy. While government repression and sometimes-horrific violence continued through the early 1990s, this broad change in strategic approach helped shift popular support away from the insurgency and toward the government. So, as Shining Path military strength increased, the group's popularity declined. By the time the security services captured Guzmán in 1992 (the first "end" of the insurgency), Shining Path had alienated a sizable portion of its rural base and made little to no headway in building support in the capital.[53] Tens of thousands of rural civilians, many of them angry and fearing further Shining Path atrocities, enlisted in government-sponsored militias.[54] These militias played a key role in undermining Shining Path tactical and information operations.

Taken together, Figures 4.6 and 4.7 represent one common path to end state for insurgencies that practice indiscriminate terror tactics: Increasing strength leads to increasing terror, which results in decreasing popularity (LTTE, Shining Path).[55] In some other cases, weakening insurgencies resort to terror, or increase their reliance on terror, out of desperation. This latter trend often plays out with splinter groups attempting to effect a last-ditch turnaround (e.g., Real IRA). When other intelligence indicators show an insurgency to be in decline, an increase in indiscriminate terror attacks may provide reinforcing evidence. McCuen (1966, p. 32) believes that weakness encourages proto-insurgencies to employ terrorism:

> Obviously, the revolutionaries will have to use terrorism in any event if they have not yet been able to develop their political and military organization sufficiently to support guerrillas. For this reason, terrorism has been called a weapon of weakness—for revolutionaries with small resources.[56]

---

[53] Shining Path's terrorist attacks almost certainly undermined the group's popularity, which, in turn, undermined its recruiting efforts and hardened the progovernment cliques.

[54] Government incentive programs also played a significant role in the militia recruiting drives. It should also be noted that government repression alienated significant segments of the population. In light of Shining Path's military momentum, it is not clear that the Fujimori government in 1992 would have been able to defeat the insurgents barring the capture of Guzmán.

[55] Some researchers believe that Shining Path conducted a selective, if brutal, terror campaign, while others saw less discipline in the insurgents' methods. For additional detail and a variety of viewpoints on the Peruvian campaign, see Stern (1998) and Palmer (1994).

[56] McCuen saw terrorism as a phase of warfare. In order, he saw the four phases as organization, terrorism, guerrilla warfare, and mobile warfare. The terrorism phase is a period of relative weakness in relation to government security forces.

Figures 4.6 and 4.7 also show notional popularity trends.[57] That terror affects public opinion differently in various regions and among various groups is not surprising. This gradation does, however, pose problems for the insurgent who must shape information operations to address varying local concerns and perspectives. Guzmán, who became increasingly dogmatic as Shining Path gained strength, proved incapable of reaching urban slum dwellers put off by his unbending Marxist propaganda. However, it was primarily his reliance on indiscriminate terror that undermined his efforts to create an urban insurrection. After Guzmán's capture, the Shining Path infrastructure quickly began to crumble, and soon, thousands of Shining Path guerrillas had either been captured or had surrendered. The "new" Shining Path insurgency (Peru failed to address the root causes of the conflict) is built on remnants of Guzmán's shattered movement.

Figure 4.6
Effect of Shining Path Terror Tactics

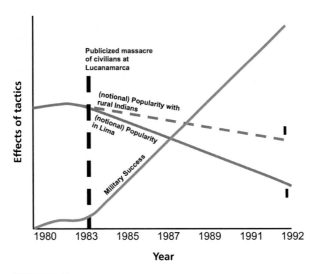

RAND MG965-4.6

---

[57] With the resources available to researchers, reliable public-opinion data on Shining Path could be uncovered only for 1992. The starting points on these lines are subjective and based on qualitative research that compared the 1992 data with descriptions of popular opinion on the insurgency in the early 1980s. The insurgency never was very popular in Lima.

**Figure 4.7
End of First Shining Path Insurgency**

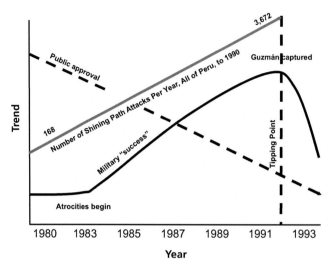

In nearly every case in which insurgent groups were defeated, the use of indiscriminate terror proved to be a considerable distraction from their primary objectives (see Table 4.3).[58] Terror often became an end unto itself as insurgents rationalized increasingly ugly brutality. This unfettered violence attracted criminals and others with no ideological ties to the movement, furthering the descent into random terror and weakening the credibility of insurgent indoctrination. This especially proved true in the case of AQI when the process of "Iraqification" devolved into mass criminalization and petty violence for personal gain. AQI's brutality backfired, and, by 2006–2007, the tribal leaders in Anbar fomented a revolt; "enough was enough for the locals" (Kilcullen, 2009, p. 173). Although AQI had not been defeated throughout Iraq by the time the research for this study concluded, grassroots revul-

---

[58] In a few cases, atrocities and indiscriminate terror proved to be useful tools for the insurgents. The Vietcong brutally executed thousands of Vietnamese civilians—often minorities, such as Montagnards—over the course of the South Vietnam conflict to good effect.

Table 4.3
Number of Insurgencies by Insurgent Use of Terror

Outcome	Indiscriminate	Mutual Atrocities	Discrete[a]	Little or None
Government wins	11	3	8	4
Mixed outcome	5	3	6	4
Government loses	5	2	14	4
Ongoing	7	4	3	1

[a] Including two cases coded as first broad and then discrete.

sion to AQI murder and intimidation tactics in Al Anbar province had denied the group its most effective internal safe haven.[59]

Insurgencies employing limited terror tactics were far more successful than those employing indiscriminate or excessive terrorism. When insurgents were able to limit their use of terror, they increased their win-loss ratio from 5:11 to 14:8 (with a number of ongoing or mixed cases). In each of these 14 successful cases, the insurgency was able to justify its actions, direct its attacks against relatively acceptable targets, or instigate the government to the point that that the population focused on the repressive tactics of the state and not insurgent terrorism (see Table 4.4).

A quotation from a prominent Naxalite insurgent leader, Nagabhushan Patnaik, reveals the cognitive dissonance that ensues when the

Table 4.4
Number and Duration of Insurgencies Sorted by Insurgent Use of Terror

Insurgencies	Broad	Atrocities	Discrete	None
Number of concluded insurgencies	21	8	32	12
Average duration (years)	8.1	12.8	15.8	8.5

---

[59] We refer here to the Awakening Movement.

movement is forced to revert to more-discrete tactics. Like counter-insurgents, insurgents periodically reassess their strategies whether they are winning or losing. In this case, terrorism backfired. In a 1981 interview, Patnaik refers to the effect that the assassinations of government officials and prominent civilians had on the first incarnation of the Naxalite movement:[60]

> Definitely it was not murder. It was punishment inflicted by the masses. Though we thought that by this we would be further-ing the cause of our struggle, it did not. So, we are correcting ourselves. In fact, we have to change our course from eliminat-ing individuals to the path of agrarian revolution. (Balakrishnan, 2004)

McCuen (1966, p. 33) summarizes the conventional wisdom of the counterinsurgents as well as our quantitative and qualitative find-ing. In the end, terrorism is

> not an efficient type of warfare. The revolutionaries cannot gain permanent support of a population by terror. Terror may, as it did in Malaya, drive people into support of the administration if the government authorities can offer them security. The wise revolutionaries will dispense with terrorism as rapidly as possible to avoid this ultimately adverse reaction.

Insurgent Strength

While insurgents are typically weaker than counterinsurgents, it is worth examining the value of military strength as a correlative or causal factor. In doing so, we developed perhaps the most subjective analysis in the study. We asked our researchers to code each insurgent group

---

[60] As of 2009, the Naxalite insurgency continued, although the term *Naxalite* has now come to generically represent a range of insurgent groups and movements operating out of West Bengal, India.

as having high, medium, or low military strength (see Figure 4.8).[61] They based their judgments on a variety of factors, including battlefield results and casualty statistics, but, in the end, they made educated judgments. With this rather weighty caveat in place, our analysis showed that not only are low- or medium-competence insurgencies capable of winning (4:4 with others ongoing or mixed), but that high-competence insurgencies are slightly more prone to failure (4:6 with others ongoing or mixed) (see Table 4.5).[62]

These findings contribute to the idea that insurgency endings are complex and that the results are often counterintuitive. Record (2007, p. 132) expresses conventional wisdom (the weak are prone to lose), while also offering a rational and applicable caveat: "Weaker-side victories are exceptional and almost always rest on some combination of stronger political will, superior strategy, and foreign help."

Figure 4.8
Insurgent Strength

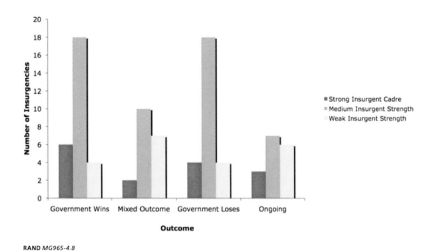

RAND MG965-4.8

---

[61] We initially coded this as "insurgent competency." However, competency speaks to leadership more than force. *Strength* better describes the capability of insurgent cadres.

[62] Ivan Arreguín-Toft (2007) provides a detailed examination of relative strength in small wars.

Table 4.5
Number of Insurgencies by Insurgent Competence

Outcome	High Insurgent Competence	Medium Insurgent Competence	Low Insurgent Competence
Government wins	6	18	4
Mixed outcome	2	10	7
Government loses	4	18	4
Ongoing	3	7	6

How can weak insurgencies win 50 percent of the time out of all decided cases? The most obvious conclusions seem also to be generally true, in concurrence with Record's caveat: The government is equally weak; the government actively loses the war through ineptitude; the root causes of the insurgency are strong enough to carry the fight to ending. Most weak insurgencies occurred in failed states or in very poorly governed states (e.g., Sierra Leone, Tajikistan, Congo, Eritrea, Ivory Coast, Dominican Republic). Government efficiency in these cases certainly correlated with insurgent victory. In fact, in each of the aforementioned cases, we also rated government military competence as "low."[63]

If both sides are weak, then we must look to other factors to fully explain the ends of these insurgencies. However, the very idea that weak insurgencies have a 50-50 chance of successfully overthrowing a government should be alarming in an era when international terrorists seek refuge in weak or failed states (e.g., Mauritania, Somalia). There is further cause for concern when this finding is coupled with that on external interventions. A force intervening on behalf of the government has less than a 50-50 chance of defeating an insurgency. This may be especially true in very poor countries, where it is almost impossible for the intervening force to address the root causes of the conflict.

Weak is also a subjective term. Here, we mean militarily weak, in that the insurgent cadres are poorly trained, poorly equipped, poorly

---

[63] Unpopularity, lack of external support, and anocratic behavior also strongly correlate with government defeat in most of these cases.

led, or all three. Their military weakness may also be relative in that they compare poorly with the counterinsurgent's forces. Military weakness does not necessarily conflate with ideological or political weakness, however. Some militarily weak insurgencies achieved political victories or forced the government to negotiate because they were able to sustain a high degree of popular support. O'Neill (1990, p. 39) draws this point out in his discussion on insurgent strategies:

> [W]hat the Algerian [independence] war showed was that victory is possible without the structured phasing and military progression associated with the implementation of Mao's strategy. The key to overcoming the deficiencies of what turned out to be more of an ad hoc approach (and military regression) was gaining and maintaining popular support through good organization and astute psychological warfare campaigns.

In cases in which highly competent insurgents lost, the government also was rated as highly or moderately competent. This proved true in Argentina and Nicaragua (Contras). When would being good at insurgency hinder the insurgent? A contest between two highly competent forces would seem to generally be a 50-50 proposition, with the government benefiting slightly from its natural advantages.

There are some ways in which high degrees of competence can correlate with defeat. Strength can lead to overconfidence and can actually help generate momentum for the COIN effort. The Sri Lankan LTTE leadership was so confident in its military capabilities that it pulled out of peace talks in 2003, even after achieving de facto semiautonomous self-rule. Six years later, Sri Lankan military forces had rebounded and crushed the LTTE insurgents. In some cases, government and military leaders (sometimes one and the same) who are unmoved by the plight of rural peasants caught up in a violent insurgency may be energized when they sense a legitimate threat to their own survival; this proved true in Greece (1945–1949), a case into which we delve in greater detail later in this chapter. O'Neill (1990, p. 48) believes that timing plays a role in the success or failure of an insurgency: If they move too soon, they risk "galvanizing major government countermeasures."

System of Government

Pseudodemocracies, or *anocracies*, have a particularly poor record at countering insurgency, winning about 15 percent of all decided contests (1:7). Autocracies win at a rate of 19:14, and democracies win at a rate of 9:0 (with other mixed and ongoing outcomes in all cases).[64] This finding strongly reinforces the idea that insurgencies do not entirely end until the government has addressed the root causes of the conflict (see Figure 4.9).[65]

This proved true even when the military wing of the insurgency had been decimated or decapitated, as we saw with the resurgence of Shining Path, the splintering of the Algerian GIA into the Salafist Group for Preaching and Combat, the resilience of the Afghan Taliban, the ongoing fighting across the southern Philippines, and in the

Figure 4.9
System of Government

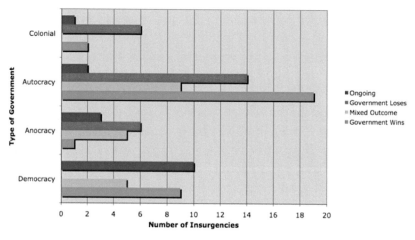

RAND *MG965-4.9*

---

[64] Statistics on democracies do not include the successes and failures of democratic sponsors supporting anocratic governments (e.g., the United States and Vietnam).

[65] Ted Robert Gurr, James Fearon, David Laitin, and James Vreeland are the definitive experts on anocracy and on the characteristics of anocratic regimes.

amorphous movement in southern Thailand, each of which survives in the wake of "defeat." Anocratic behavior by each of the governments in question—Peru, Algeria, Afghanistan/Pakistan, the Philippines, and Thailand—sustained and stoked the kind of grassroots discontent necessary to rekindle an insurgency.

Conventional wisdom on both COIN and insurgency also points to the necessity of addressing a conflict's root causes. The body of professional COIN literature clearly identifies that COIN is essentially a struggle for the hearts and minds of the population. The counterinsurgent must provide the population security, a stable economy, and (legitimate) basic rights if the COIN is to succeed. Galula (1964 [2006], p. 102) believes that even a half-hearted effort is worthwhile:

> [Knowing] that his program will have no or little immediate appeal, the counterinsurgent must somehow find a set of reforms, even if secondary, even if minor. He has to gamble that reason, in the long run, will prevail over passion.

To varying degrees, each of the insurgent philosophers we have identified exhorts similar themes: Government reform threatens critical grassroots support for insurgencies and should be undermined at all costs. So, in theory, government reform may be the best tool with which to defeat an insurgency. Practical application of this theory, however, is decidedly mixed. Anocracies are inherently weak in that they are ineffective at employing both democratic and autocratic methods. These governments have only few options to escape their fate: by fully democratizing, by crushing the insurgents and repressing the populace, or simply by trying to ride out the storm in the hope that other factors lead them to victory. With these options in mind, there essentially are four ways in which anocratic counterinsurgencies end:

- Case 1: A government starts and ends as an anocracy with no real effort to change its behavior, and typically loses (e.g., South Africa).

- Case 2: It tries (often half-heartedly) and fails to democratize during the conflict and is subsequently defeated (e.g., South Vietnam).[66]
- Case 3: An anocracy successfully democratizes or recognizes minority rights and is able to realize a favorable ending (e.g., Northern Ireland).
- Case 4: An anocracy slips into autocracy to survive (e.g., Algeria GIA).

Each of these outcomes contains lessons. The first case also is the simplest: An anocracy that makes little or no attempt to democratize is likely to fail over time. Because it never addresses the central complaints that inspire the insurgency, the government helps sustain the fundamental bond between the insurgent and the population.[67] And, in its efforts to appear democratic, the regime forgoes the use of ruthless repressive tactics, tactics that have proven effective (if often temporarily) in stifling some insurgencies. The anocracy is weak on all fronts and susceptible to defeat. Insurgent propagandists excel at exploiting the inherent fallacies of anocratic governance, while insurgent cadres are more likely to thrive in the absence of acute repression. External support that could tip the balance in other cases is less relevant, or irrelevant, when the government makes little effort to help itself. Anocracies that do not change are often soundly defeated.

The second case is linked to an ongoing debate among some COIN and political theorists: Are the risks of democratization outweighed by the gains? An anocracy finds itself in a "damned if you do, damned if you don't" situation.[68] If it tries to democratize after an insurgency has erupted, it can actually endanger its own stability and survival. McCuen (1966, p. 59) believes that timing is critical: "Grad-

---

[66] This option includes the possibility that a government successfully democratizes and is still defeated militarily or by means external to a legitimate political process. To "succeed," reforms must also be accepted to the point that they undermine the insurgency.

[67] Recalling Mao's thesis, success may require support from only 15–25 percent of the population.

[68] Paul Collier (2009) breaks down this Faustian bargain in detail in *Wars, Guns, and Votes: Democracy in Dangerous Places*.

ual reforms initiated early usually will eliminate the need for rapid, radical ones later on."

An anocratic government would be naturally hesitant to take on Galula's "gamble." As seen in Figure 4.10, in some cases, acquiescence to political demands can lead to a cascading series of strategic disasters. While granting freedom of the press demonstrates a recognition of basic rights, it also allows the press to condemn government reforms as half-measures, criticize military tactics and leadership, and call for further, and further damaging, transformation. Freedom of speech allows supporters of the insurgency to stage mass rallies against the government, which might then develop into a general revolt. Freedom from random search and seizure handcuffs an aggressive security service. Speedy trials deny the use of emergency-power detentions, and freedom of religion may simply be too much to contemplate for nominal theocracies.

Acquiescence to demands for targeted, group-specific rights can be equally toxic to long-term stability. When the governments of Colombia and Sri Lanka granted the FARC and LTTE (respectively)

Figure 4.10
Notional Endings of Anocracy Cases

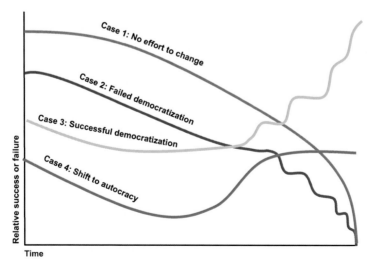

Case 1: No effort to change

Case 2: Failed democratization

Case 3: Successful democratization

Case 4: Shift to autocracy

Relative success or failure

Time

semiautonomous zones, both groups simply used the zones as safe havens and continued to build strength and fight.[69] Both governments were eventually forced to mount extensive military campaigns in an effort to root out the insurgents. When governments do attempt to grant various rights, they often do so in half-measures and only under tremendous pressure. In this way, it appears that they are conceding to insurgent demands, a sign of weakness that simply reinforces the impression that the insurgency is winning. Unrelenting insurgent propagandists jump at the chance to embarrass the government, which, in its weakened state, may offer added reforms. It quickly finds itself being dragged down by its own inconsistencies.[70]

The third case is the best-case scenario for the counterinsurgent; it is also very rare. According to Hammes (2006, p. 82), Sandinista leader Daniel Ortega clearly understood the danger of successful democratization to an insurgency movement:

> The greatest threat to the Sandinistas was the democratic reform movement. If it succeeded while the Sandinistas were still getting established in the mountains, the cause of the insurgency would be neutralized. The formation of a moderate, reformist government would present the Sandinistas with a much greater problem than a repressive Somoza regime.

Qualitative analysis of the 15 identified anocracies showed that, once fighting began, it proved extremely difficult for the government to address root social causes while also retaining the reins of power. In many cases, the best an anocracy could hope for was a negotiated settlement that resulted in power sharing or autonomy for a repressed minority (e.g., El Salvador 1979–1992 and Bosnia 1992–1995). In

---

[69] We counted Colombia as a democracy, not an anocracy. Vreeland (2008) points out that Sri Lanka was probably a functioning democracy until 1983, the point at which we assess that the LTTE insurgency began in earnest. In this case, the government slipped from democracy into anocracy in response to the insurgent political and military threat.

[70] Although we did not study the fall of the shah of Iran in 1979 as an insurgency case, the shah's last-ditch efforts at reform and his subsequent defeat exemplify the spiraling effect of too-little, too-late reform.

only one case did we assess that an anocracy had achieved genuine and *successful* democratization (Croatia 1992–1995), and even this case is debatable on a number of fronts. It is important to note that genuine reform and success are not necessarily coterminous. Case 2 clearly shows that genuine reform can and does fail. How, then, can anocracies win?

Data on successful COIN campaigns conducted by anocracies are thin.[71] However, our broader analysis of the full range of 89 cases leads us to a middle ground in terms of conventional wisdom: Security and democratization (or at least electoral freedom and basic human rights) are not only mutually supporting elements of COIN, but also they must both be present to establish a firm, secure end state. In the following few paragraphs, we attempt to present the homogenized viewpoints of Galula, Merom, Collier, Weinstein, Hammes, Petraeus, and others.

To succeed, the counterinsurgent should attempt to establish and preserve security, minimize or eliminate public opposition to its policies through genuine and lasting reform, ensure the availability of essential services and employment, and either destroy the insurgents or co-opt them in a way that does not risk the regime's authority or legitimacy.[72] Establishing security requires all the military and police functions necessary in any COIN campaign, but anocracies must conduct security operations in a way that does not delegitimize their reform efforts. Security forces must be especially careful to abide by both national laws and, depending on the situation, international laws of land warfare. Further, they must also understand and navigate local customs.[73] Security forces cannot conduct operations in a vacuum: When they cause damage, they must be prepared to pay recompense and follow up

---

[71] Here, we define success as a reduction in violence to peacetime levels, grassroots grievances redressed, all groups operating within a defined and accepted political process, and external forces either out of the country or far along in the process of relinquishing authority to local law enforcement.

[72] Another obvious alternative is case 4, which we describe later in this section.

[73] In many of the successful COIN operations we studied, including Malaysia, Iraq (so far), Uruguay, and Colombia, well-trained and professional local police and local militia proved most adept at navigating cultural obstacles and were critical in the long-term success of the campaign.

with civil aid. In some cases, external sponsors secured the population long enough for the government to democratize; additional forces and monies can speed security efforts considerably. Of course, according to our findings, this support has less than a 50-50 chance of giving any government a winning edge.

Public support can be won only if reforms are both legitimate and effective. It is not enough for the government to mean well—or pretend to mean well—it must also follow through consistently over a period of time. The government must actually transform into at least a rudimentary representative government. As McCuen (1966, p. 59) puts it,

> [Effective] persuasion requires concrete evidence of action or implementation at the local level. The revolutionaries back up their glowing promises for the future with such popular deeds as land reform, elimination of usurious interest, killing of hated officials, and so forth. Although the government power will encounter opposition within its own ranks to such actions as land reform, reduction of interest, and removing of disliked officials, these should be accomplished to the extent necessary to win popular support.

Civilians suffering through an insurgency will be slow to trust either side and will be very cautious in throwing their support behind the government. Support also may depend on the establishment of basic services, such as adequate food, clean water, electricity, and employment. It seems obvious that both services and basic rights should be addressed simultaneously, and, indeed, this is a best-case scenario. However, anocracies in the throes of a violent insurgency are often hard-pressed to make even incremental improvements in governance.

Barring independent collapse, defeating an insurgent military cadre requires either total elimination of its leadership and infrastructure or co-option. Co-option has proven to be the better of the two options. Annihilation of an insurgent cadre typically requires massive expenditures in money, broad use of force that commonly leads to civilian casualties, and (if it has passed the protoinsurgency phase) time. The counterinsurgent would be better off balancing security with reconstruction and reconciliation programs and avoiding civilian casu-alties in an effort to win hearts and minds; time is never a guaran-teed commodity. Co-opted insurgents lend legitimacy to the political process and often contribute positively to the development of genuine reform (e.g., Colombia FARC).

That Iraq in mid-2009 was an anocracy is probably a contest-able issue, but we treat it as such. We offer that Iraq rests somewhere between cases 2 and 3: It is attempting to democratize, but its path has not yet been determined. At the very least, U.S. campaign design post-2007 incorporates many of the elements we identify as keys to success. With U.S. guidance, the Iraqi government is slowly work-ing to negotiate social reforms; the provincial election in early 2009 was universally judged as legitimate; security operations are designed and increasingly executed with cultural and legal limitations in mind; reconstruction and reconciliation efforts are accelerating (albeit in fits and starts); and tens of thousands of Sunni who made up the heart of the anticoalition insurgency have been co-opted into lawful militias. U.S. forces continue to act in a stabilizing role, giving the Iraqis time to reconcile their differences. From this point forward, two questions will determine Iraq's future: Will the process of democratization con-tinue without dominant U.S. influence over Iraqi national policy? And are the myriad underlying social issues reconcilable without civil war or regression into dictatorship?

Anocracies have a final choice, of course. They may abandon most, if not all, pretenses of democracy and human rights and implement a range of repressive, autocratic policies, a course not available to "true" democracies. In fact, Merom (2003, p. 15) states that democracies are incapable of the absolute repression sometimes necessary for survival:

> My argument is that democracies fail in small wars because they find it extremely difficult to escalate the level of violence and bru-tality to that which can secure victory.

One could also say that, in extremis, a democracy (or, in this case, an anocracy) has a choice: It can abandon human rights in order to survive, or it may perish. As insurgencies expand and threaten the central government, the temptation to resort to the mailed fist becomes overpowering. Survival instincts take hold, military advisers rationalize the use of emergency authority laws, and, within a short period of time, the anocracy has transformed itself into (or been replaced by) a dictatorship or near-dictatorship (e.g., Algeria). Statistically, this is the second-best survival option, next to successful democratization, without many of the risks inherent in that process. There is little chance that an anocracy will fail to turn into an autocracy and, based on our findings, probably better than a 50-50 chance of success once it does. In most cases, however, anocracies probably only forestall their own doom by resorting to autocracy, since they fail to address any of the root causes of the insurgency and often worsen conditions along the way.[74] Merom (2003, p. 22) is not alone in pointing out that insurgents exploit the excesses of autocracies, stating that they "also try to lure democratic opponents into behaving brutally in order to increase the moral opposition to the war."[75]

Figure 4.10 depicts notional endings for each of these four cases. For case 1, the anocracy slides with increasing rapidity into defeat, in accordance with McCormick's findings. In case 2, the ripples at the end of the curve represent an increasingly rapid cycle of reform and setback that cascades into eventual defeat. Case 3 reflects the same rippled reforms, but, in each situation, the reforms are generally successful. In the final case, the autocracy survives but is perpetually burdened with low-level violence and civil unrest (hence the lower ending point).

A number of the 89 cases help describe the "shift to autocracy" (case 4). The Algerian government's behavior in response to Islamic political victories in the early 1990s provides one of the best examples (see Figures 4.11 and 4.12).

In looking at all 89 cases, but in particular the 15 cases of anocracy, we found that, by properly timing democratization, reform, and reconciliation, the counterinsurgent could effect a tipping point and

---

[74] Some regimes, of course, are lucky enough to survive the inevitable, passing along the specter of defeat to their replacements.

[75] He includes an excellent branching diagram (2003, p. 23) that depicts this kind of insurgent strategy. Also see McCuen (1966, p. 61).

**Figure 4.11
Algerian Anocracy**

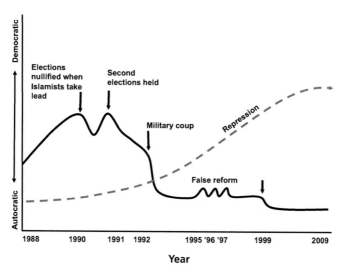

RAND *MG965-4.11*

**Figure 4.12
Outcome of Algerian Insurgency**

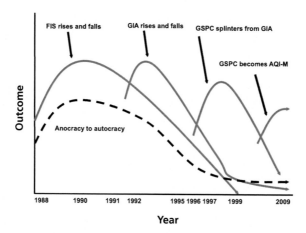

NOTE: FLS = Front de Libération Nationale, or National
Liberation Front. GSPC = Groupe Salafiste pour la
Prédication et le Combat, or Salafist Group for Preaching
and Combat. AQIM = al Qaeda in the Islamic Maghreb.
RAND *MG965-4.12*

shape a favorable outcome. In the case of Algeria, the government made a few rather desultory and (probably) disingenuous efforts to democratize in the throes of the GIA insurgency, before security had been established, and then failed to take any serious overt steps toward democratization as the GIA faded in the late 1990s. *Across the board, reform in the absence of security proved either ineffective or calamitous.* In Iraq, it proved possible to bull through periods of poor security and half-hearted reform, but this "bulling through" came at a widely publicized cost. Kilcullen (2009), Hammes (2006), and Galula (1964 [2006]) all concur that security has primacy. FM 3-24 (p. 1-3) neatly breaks a COIN campaign into stages:

> Gaining and retaining the initiative requires counterinsurgents to address the insurgency's causes through stability operations as well. This initially involves securing and controlling the local populace and *providing for essential services* [emphasis ours]. As security improves, military resources contribute to supporting government reforms and reconstruction projects. As counterinsurgents gain the initiative, offensive operations focus on eliminating the insurgent cadre, while defensive operations focus on protecting the populace and infrastructure from direct attacks. As counterinsurgents establish military ascendancy, stability operations expand across the area of operations (AO) and eventually predominate. Victory is achieved when the populace consents to the government's legitimacy and stops actively and passively supporting the insurgency.[76]

Galula (1964 [2006], p. 120) also divides COIN into sequenced phases, and he too warns against attempting reform during the pacification phase of a COIN operation:

---

[76] We emphasize the words *providing for essential services*. This implies that the military or civil-military units act as suppliers of food, water, and medical aid in the earliest stages of a campaign—a reasonable first step. However, this step should not be equated with the kind of reform necessary to democratize. The Marine Corps' (1940, p. 5) *Small Wars Manual* identifies five phases.

Implementing political reforms—if they have been conceived and announced by the government—would be premature at this stage. The time will be right when the insurgent political cells have been destroyed and when local leaders have emerged.

We observed exceptions or potential exceptions to these principles in many of the 89 cases (see Table 4.6). As with all things COIN, strategy and tactics are situation-dependent. *If it does not come at the expense of security*, phased reform can and should be instituted contiguously with other measures in specific geographic areas. For example, as of late 2009, the progress of reform in Kabul, Afghanistan, outpaced progress along the border with Pakistan, while, in Iraq, progress in both Najaf and Basra—both relatively secure by mid-2009—outpaces progress in Mosul. Andrew Krepinevich (2005b), one of our noted COIN scholars, describes asynchronous, geographically targeted actions like this as "oil spotting," a strategy that has been adopted at least in part in both Iraq and Afghanistan. Asynchronous political reform in a targeted area could come in the form of local elections, reduced emergency powers, or other reforms enacted by local, rather than national, governments.[77]

Table 4.6
Number of Insurgencies by Type of Government

Outcome	Democracy	Anocracy	Autocracy	Colonial
Government wins	9	1	19	2
Mixed outcome	5	5	9	0
Government loses	0	6	14	6
Ongoing	10	3	2	1

[77] McCuen and many other authors have also used the term *oil spotting* in a COIN context.

CHAPTER FIVE

Assessments of Insurgency Endings: Other Factors

This chapter addresses topics that we did not distinctly address in the quantitative portion of the study but found to be qualitatively critical to understanding insurgency endings.

Force Ratios

We originally approached the topic of force ratios by comparing the proportion of counterinsurgents to insurgents.[1] Our finding showed that a dominating force presence (9:1 or greater) correlated strongly with success for the counterinsurgent and that taking on an insurgency with a 1:1 (or even 2:1) force ratio was imprudent. One might assume from this finding that more forces are better. However, we relegated this finding to the appendixes because it is vulnerable to a range of valid criticisms: Force sizes tend to shift significantly on both sides over the course of an insurgency; accurately counting insurgents is *at best* a dubious undertaking and at worst impossible;[2] and, perhaps most importantly, conventional wisdom focuses on troop-to-population,

---

[1] Some researchers refer to a *force ratio* as the ratio between counterinsurgent and insurgent and *force density* as the ratio between counterinsurgent and population. We use *ratio* for both.

[2] Contemporary researchers are divided on the notion of counting insurgents. One of the authors of *How Insurgencies End* has direct experience attempting to count insurgents in Iraq and participated in a high-level U.S. intelligence panel that determined that it was not possible to accurately count Iraqi insurgents.

not troop-to-insurgent, ratios. Because COIN is a population-centric endeavor, this last criticism is also the most relevant. We therefore briefly address the conventional wisdom on troop-to-insurgent ratios and then, in the interests of exploring conventional wisdom, enter into a more detailed discussion of troop-to-population ratios.

Assuming that one can accurately count insurgents, or at least insurgent combatants, force-on-force calculations might take on greater significance than they would in a typical COIN setting. A positive, albeit temporary, end state to the conflict could theoretically be reached through simple attrition. However, even using troop-to-insurgent math, the numbers are probably insufficient to guarantee success. Here, Jeffrey Record (2007, pp. 58–59) describes French troop-to-insurgent ratios in Algeria:

> French forces in Algeria peaked at 500,000, including at least 200,000 mobilized reservists, in 1960, by which time there were only a few thousand (insurgents) still operating inside Algeria. French forces always maintained at least a 10:1 numerical advantage over the insurgency inside Algeria.

While various French and British COIN experts (Galula, among others) emphasize population-centric COIN warfare, the British army pays brief homage to force ratios and, coincidentally, to attrition warfare:

> It would be wrong to deduce that any application of attrition is necessarily counter-productive: in Malaya the British were able to achieve a force ratio of 20:1,000 and used their military superiority in numbers and firepower as a means to drive Chin Peng's communists into remote parts of the country, where they were then hunted down remorselessly. (UK Ministry of Defence, 2001, p. B-2-2)[3]

---

[3] It should be noted that the British were facing a geographically and ethnically isolated insurgency and (in the latter stage of the war) were simultaneously conducting an otherwise sound COIN campaign.

Neither Record nor the British army implies that force alone can win a COIN fight. We have shown that even the defeat of an insurgent cadre does not end the insurgency in the absence of social rapprochement. Clutterbuck (1966, pp. 42–43) believes that attempts to define troop-to-insurgent tie-down ratios are both "nonsense" and a "dangerous illusion."[4] Most experts and doctrinal publications instead address troop-to-population ratios. See, for example, FM 324 (p. 1-13):

> No force level guarantees victory for either side. . . . [N]o predetermined, fixed ratio of friendly troops to enemy combatants ensures success in COIN. The . . . operational environment . . . and approaches insurgents use vary too widely.

If sufficient troops are available to secure the population, then social reform and combat operations can be conducted in parallel to bring the campaign to a successful conclusion. Conventional wisdom places the optimum ratio at 20 security-force personnel to 1,000 civilians (20:1,000). For example, 20,000 troops—both internally and externally sourced—and police would be sufficient to secure a country of 1 million people.

RAND researcher James T. Quinlivan's 1995 article in *Parameters* is the most oft-cited article on the subject of troop-to-population ratios. Quinlivan (1995) identified a 20:1,000 ratio as optimal, or at least minimal, to win a COIN fight. His finding has been broadly accepted by planners and policymakers and has become the de facto marker for conventional wisdom on the subject. Figure 5.1 comes from a 2003 article Quinlivan published in *RAND Review.*

Peter J. T. Krause (2007, p. 2) of the Massachusetts Institute of Technology sees value in Quinlivan's analysis:

> Quinlivan's article sits at the nexus of policy and science, and so carries some of the strengths and weaknesses of each. Policymakers need answers to clear, relevant questions and they need them now. A 50 percent reliable answer today is often far more valuable

---

[4] Both James T. Quinlivan (1995) and Peter J. P. Krause (2007) also cite Clutterbuck on ratios.

Figure 5.1
Quinlivan Force Ratio

Successful Nation-Building Usually Requires 20 Troops per Thousand Population[a]

[a]Maximum international forces for all countries (except Iraq) taken from *America's Role in Nation-Building*, Dobbins et al. Totals are as follows: Somalia, 28,000; Haiti, 23,000; Bosnia, 60,000; Kosovo, 45,000; Afghanistan, 14,000. Current total of 150,000 for Iraq is based on latest news reports.
[b]1993 population of Somalia: 6,059,950. Source: U.S. Bureau of the Census, International Data Base.
[c]1994 population of Haiti: 6,500,213. Source: U.S. Bureau of the Census, International Data Base.
[d]1996 population of Bosnia: 2,656,000. Source: U.S. Bureau of the Census, Report WP/96, World Population Profile: 1996, Washington, D.C.: Government Printing Office, 1996.
[e]1999 population of Kosovo: 1,900,000. Source: UNMIK Kosovo Fact Sheet (cites a population figure for Kosovo in 2002 of 1.7–1.9 million).
[f]2002 population of Afghanistan: 27,755,775. Source: U.S. Bureau of the Census, International Data Base.
[g]2003 population of Iraq: 24,683,313. Source: U.S. Bureau of the Census, International Data Base.

RAND *MG965-5.1*

than a 95 percent reliable answer two years (or even two weeks) from now. Quinlivan's piece provides a clear answer to the question: how many troops do we need to seriously consider undertaking a given stability operation? The 20 per 1,000 ratio isn't perfect, but it has proven a decent ballpark figure.

However, Krause (2007, pp. 2–3) goes on to express concern regarding the methodology behind the study:

From a social science perspective, however, his study reveals significant flaws. Quinlivan's lack of methodological clarity concerning key terms and hypotheses inhibit rigorous testing of his claims. . . . Even the cases he examines pose problems for his argument, since only two represent stabilization successes with ratios of 20 troops per 1,000 (Malaysia and Northern Ireland) while others achieved stability with ratios in the single digits (Germany following the Second World War, India in the Punjab in the mid-1990s, and the U.S. in the Dominican Republic in 1965). Further examination reveals cases with intervention forces yielding troop ratios above 20 per 1,000 that were unable to maintain stability, such as the French in Algeria.

John J. McGrath conducted in-depth analysis on troop-to-population ratios as recently as 2006. In *Boots on the Ground: Troop Density in Contingency Operations*, McGrath (2006) coded a range of insurgency data, including geographical area, terrain, population density, troop deployment and organization, and indigenous support. He found a ratio of 13.26 security-force personnel (including police) for every 1,000 citizens to have been sufficient to conduct successful COIN operations in historical cases. While McGrath delved deeper into a wider range of case studies than Quinlivan, the two methodologies were, in many ways, similar.

Most COIN theorists and experts discuss the value of troop-to-population ratios through historical case studies.[5] A few equate terrain with both population and insurgents. Record (2007, pp. 54, 81) believes that the low density of Soviet forces in Afghanistan correlated with their defeat, comparing their troop levels to U.S. troop levels in Vietnam. The Soviets peaked at 90,000–120,000 troops in a country five times as large as Vietnam, where the United States fielded more than 500,000 troops circa 1968–1969:

> The Soviets committed the cardinal strategic sin of committing *under*whelming force—a sin the United States seems to have later repeated in Iraq [where] it is clear that U.S. forces have not been sufficiently strong either to protect the threatened populations or to impose on the insurgency a combat loss rate beyond its ability to replace.

---

[5] The Institute for Defense Analysis, Center for Army Analysis, and the British Defence Science and Technology Laboratory conducted case-study analyses of force ratio in 2009. The Center for Army Analysis also conducted a troop-to-insurgent study. The Center for Naval Analysis conducted an in-depth, subdistrict study of force ratios in Afghanistan. None of these studies was available for citation at the time *How Insurgencies End* went to publication.

Hammes (2006, p. 186) believed that U.S. troop levels were insufficient in Iraq at the time *The Sling and the Stone* was published. He compares the force ratio in Iraq to previous U.S. operations:

> The ratio of coalition forces to civilian population in Iraq is a fraction of that for forces initially employed in Bosnia and Kosovo. During the first year of operations in Bosnia and Kosovo, there were roughly nineteen troops per thousand inhabitants. In Iraq, the ratio is fewer than seven per one thousand inhabitants. Clearly, we have too few troops to provide security and nation-building assistance. . . . One response to the shortage of troops appears to be a rush to count hastily trained Iraqis as qualified security personnel.

By comparing a relatively nonkinetic deployment like the U.S. mission in the Balkans to the war in Iraq, Hammes seems to compare a rather nonkinetic peacekeeping operation with an all-out COIN war. Hammes is not an outlier: Most COIN studies draw from a broad spectrum of conflicts. Debate over the fine line between COIN operations and stability operations is commonplace but also subjective. Taking a broad approach to case-study coding, a 2005 RAND study on the subject of force ratios in stability operations may also be relevant to COIN. The authors of that monograph (Jones et al., 2005, p. 19) saw a potential requirement for 1,000 security-force personnel for every 100,000 civilians but also show that ratios varied widely from operation to operation:

> There were more than 10,000 U.S. troops per 100,000 inhabitants in the American sector of Germany after World War II, 1,900 troops (per 100,000) in Bosnia, 3,400 in Eastern Slavonia, 2,000 in Kosovo, and 1,100 in East Timor in the first years of reconstruction. . . . Ratios of less than 500 troops per 100,000 inhabitants were sufficient to successfully stabilize Japan after World War II, Namibia in 1989, El Salvador in 1991, and Mozambique in 1993.

Both the Jones et al. study and most other COIN experts discuss the value of police both in meeting security-force ratios and in shaping endings. The police can play a significant role from the beginning but are especially valuable as law and order begin to take hold and they can be protected from most direct attacks. McCuen (1966, p. 205) sees the police as an excellent economy-of-force measure. Recommended police-to-civilian ratios can vary from 150 per 100,000 to 200 per 100,000 (Jones et al., 2005, p. 19).

Some experts, including Kilcullen (2009, p. 184), believe that the idea of troop ratios is overblown and irrelevant. Instead, they focus on the way the troops are employed and on their success in securing and winning over the population:

> Merely adding additional foreign troops cannot compensate for lack of local popular support—the British lost the Cyprus campaign with a force [troop-to-insurgent] ratio of 110 to 1 in their favor, while in the same decade the Indonesians defeated Dar'ul Islam with a force ratio that never exceeded 3 to 1, by building partnerships with communities and employing them as village neighborhood watch groups, in cordon tasks, and in support function.

Both Kilcullen and Carter Malkasian point out that, in large, rural countries, it may be impossible to deploy sufficient forces to meet the 20:1,000 ratio. In his article on troop ratios in Anbar province, Iraq, Malkasian (2007, p. 121) points out costs of trying to meet this requirement:

> The ratio of 20 security personnel per 1,000 civilians far exceeds what the combined coalition and Iraqi indigenous forces could provide in Al Anbar. Thirty-one U.S. battalions would have been required to attain this ratio in just the 11 cities [in the region].

In formal doctrine, the U.S. Army (FM 3-24, p. 1-13) appears to reference Quinlivan but caveats its analysis of the subject:

> Most density recommendations fall within a range of 20 to 25 counterinsurgents for every 1000 residents in an AO. Twenty counterinsurgents per 1000 residents is often considered the minimum troop density required for effective COIN operations;

however as with any fixed ratio, such calculations remain very dependent upon the situation.

We could find only one published study (Quinlivan, 2003) that put the recommended ratio at or above 20:1,000. Further, the addition of the phrase "in the AO" greatly complicates troop-to-population analysis. This means that one can count only those troops that are physically located within a designated conflict zone. For example, the ongoing Thai insurgency is generally isolated in southern Thailand, so troops conducting noncombat operations in the far north would not be counted. For any insurgency case, in the absence of precise deployment data, over time, it would be nearly impossible to clearly identify force levels within any specific AO.

In fact, a key problem with COIN data is that there may be too many inputs and unknowns to generate a valid basis for calculation. While it may be possible to develop a mathematical model that accurately relates historical troop levels, historical population levels, and reported levels of violence, it is less easy to determine the accuracy or relevance of the data that would be used to build the study. Iraq offers a specific example.

The Brookings Institute's Iraq Index reported the existence of 205,700 Iraqi security forces of various types on duty in March 2004 (Brookings Institute, 2009). When added to the international troop level of approximately 140,000 in Iraq and 50,000 in Kuwait, the coalition could field nearly 400,000 security forces. However, a closer look at these numbers reveals flaws that may be inherent in any troop-ratio calculation:

- In March 2004, nearly 74,000 of the identified Iraqi security forces worked for the Facilities Protection Service (FPS) providing little more, and often less, than the security provided by a poorly paid night watchman. While some served capably, many did not show up for work or were paid off by insurgents or criminals. This force was generally considered ineffective, and it is unclear whether many—possibly thousands—of FPS guards ever showed up for work.[6]

- A sizable portion of the police force most likely moonlighted with the insurgency, committed ethnic atrocities, or directly supported insurgent activities. Hundreds or thousands of police were "ghosts" who existed only on paper so superiors could take their salaries.[7]

- Several Iraqi army units proved incapable even of engaging insurgents in combat in April 2004. The army had no logistics capability of which to speak and could not subsist without direct U.S. support.

- The 40,000-strong Iraqi Civil Defense Corps occasionally conducted massed patrols that were, in effect, little more than shows of force; they conducted little to no COIN work. They most often stayed in barracks, and many did not show up for work. Coalition forces disbanded this organization within a year.

- Brookings reported that, even by 2005 (after another year of development), most Iraqi security forces were "partially capable of conducting counterinsurgency operations in conjunction with Coalition units." This rating put the bulk of Iraqi security forces one level above a rating that generally equates to "incapable."

- While the coalition had 190,000 troops in theater, it is unclear whether one should count the 50,000 troops in Kuwait. Are these troops in the AO? Should they be discounted in an effort to take into account troop-to-tail ratios, a concern Quinlivan (1995) addresses in detail? Should support forces in Qatar or other

---

[6] The Iraq Study Group report stated that the FPS "units have questionable loyalties and capabilities. . . . One senior U.S. official described the Facilities Protection Service as 'incompetent, dysfunctional, or subversive.' Several Iraqis simply referred to them as militias" (Baker, Hamilton, and Eagleburger, 2006, p. 14).

[7] Even as of late 2006, the police were considered corrupt and ineffective:

> Iraqi police cannot control crime, and they routinely engage in sectarian violence, including the unnecessary detention, torture, and targeted execution of Sunni Arab civilians. The police are organized under the Ministry of the Interior, which is confronted by corruption and militia infiltration and lacks control over police in the provinces. (Baker, Hamilton, and Eagleburger, 2006, p. 13)

regional countries count toward force totals? It is not clear that one could accurately apply a rule regarding AO across the range of mismatched cases found in a typical force-ratio study.

- The quality and capabilities of international forces varied considerably; it is not readily apparent that all troops—in any case study we researched—are equal. If this is true, then an accurate analysis must also compare relative capabilities both within individual cases and across cases, dramatically compounding analytic complexities. Can one equate an FPS guard with a U.S. Army or Marine infantryman in a total force calculation? Can Iraqi police circa 2004 be equated with Malayan police circa 1953, or even Iraqi police circa 2009?

- Further complicating the troop-to-population math, one must note that Iraq last conducted an official census in 1997. Assuming that the numbers acquired and provided by the Saddam Hussein government were accurate, any extrapolation of these data past 2003 would be wildly skewed by wartime displacements and deaths. For example, according to the United Nations High Commissioner for Refugees (2007), up to 4 million Iraqis— approximately 15 percent of the official prewar population— might have been displaced by 2007. These 2007 data, too, are questionable because they are a gross approximation drawn from many different sources of varying levels of reliability.

For a variety of reasons, this combined 2004 troop level—more than 100,000 below the 20:1,000 ratio, if official population figures were accurate—failed to secure Iraq. Violence continued apace and intensified in 2006. The surge in 2007 brought the coalition and Iraqi security force total to 610,000, a number that exceeded Quinlivan's 20:1,000 ratio. Violence fell dramatically. However, over the same time period, nearly 100,000 Sunni Iraqis, many of whom were insurgents, joined anti-insurgent militia. Should these militiamen be counted in the force-ratio calculations? If force-ratio calculations have to account for militia forces, then the perhaps hundreds of thousands of progovernment militia members (Home Guard) enlisted by the British in

Malaya would dramatically throw off the original baseline for the 20:1,000 ratio.

In April 2009, the coalition fielded a combined force of 787,000 in Iraq. This total puts more than 200,000 forces in the field in excess of the 20:1,000 ratio, not counting tens of thousands of legitimized progovernment militia members. It is unclear whether this number is sufficient, whether the 610,000 troops in 2007 were sufficient, whether the militia forces made a tremendous difference, or whether a higher-quality force of 400,000 in 2004 would have done the job. It is also unclear whether any of these numbers are relevant in the absence of census data that may not be even remotely accurate.

The factor of time further skews the data sets for troop-ratio research. For example, when future research refers to Iraq as a finished case study, which numbers should be attached to the case? Should it be the highest troop-to-population ratio at any point? What if this ratio is achieved well after a tipping point has been reached? For example, if Iraq reached a theoretical tipping point in 2007, should researchers use the 2007 ratio, or should they use a number further along the campaign arc? Considering that there are few agreed-upon tipping points in many cases and no visible tipping point in many others, how could researchers determine that a force level is sufficient? The gradual, tailing nature of government victories we identify in our study compounds the timing problem.

How would time affect the study of a specific case like Vietnam? Should researchers tie the 1968 troop ratio (also the high-water mark for U.S. forces, at approximately 535,000) to a theoretical tipping point, or should they use the 1973 troop level—the point at which the collapse of South Vietnam began to accelerate—of a few hundred advisers? Or, should researchers aggregate the full range of troop-level data? If so, what could one ascertain by averaging the 1962, 1968, and 1973 troop levels in Vietnam? In addition to the South Vietnamese military forces (another layer of complexity), an array of irregular South Vietnamese forces would have to be included in the calculation or discarded (e.g., Montagnard militias, Popular Force units, thousands of Chieu Hoi defectors). All of these timeline considerations and others affect each case embedded within quantitative case studies. We also found that each insurgency data set available to researchers relied on slightly or significantly different methods to code these numbers, undermining comparative analysis between research models.

Researchers and policy advisers must also consider the varying degree of official detail available from case to case. Military staffs and academic researchers have monitored, tracked, and recorded the details of coalition operations in Iraq and Afghanistan with an unprecedented degree of focus and attention. Data on many historical cases either do not exist in comparable quantity or suffer from similar or often greater inaccuracies.[8] For example, while hundreds (if not thousands) of government and independent researchers kept sometimes excruciatingly detailed records on operations in Iraq and Afghanistan, there is less information available to Western researchers about Soviet and Cuban COIN operations in Angola, an oft-cited case. Are Cuban troop deployment numbers and timelines in the mid-1970s accurate? Were they properly recorded, or were they perhaps manipulated by communist functionaries? What were the Cuban troop-to-tail ratios in comparison to U.S. troop-to-tail ratios in Vietnam? Is it even remotely possible to determine how many Cuban troops were in a specified AO at any one time? Peruvian data on COIN operations against Shining Path are suspect for a host of reasons, including corruption and lack of transparency on both sides. It is not clear that, in the early 1980s, the Guatemalan army either kept accurate records or had the collection assets on hand to obtain relevant data on the population. It would be imprudent to assume that either Castro or Batista kept or provided accurate data during the Cuban insurgency. No matter how these numbers are studied or simplified, the myriad inconsistencies that plagued the research for *How Insurgencies End* also shape the reliability of troop-ratio research.

Krause correctly points out that, despite these unavoidable pitfalls, at some point along the campaign timeline, policymakers and staff planners have to consider force ratios, or at least force levels. By

---

[8]   Many researchers have found that better data exist in cases with direct foreign intervention, probably due to additional media exposure and analysis by sponsor governments.

relying on mathematical ratios drawn from historical case studies, however, they risk committing an ecological fallacy. The statement

> A force ratio of 20:1,000 has equated with success in historical case studies. Therefore, this same ratio should be sufficient to achieve victory in an upcoming or ongoing operation.

is scientifically unsound. And, because predictive modeling for force ratios focuses on national-level data, it can also suffer from data paradoxes like Simpson's paradox: Subsets of data can appear to provide positive results at one level of analysis, but negative results at another. This paradox can occur when subsets of variable data are aggregated into one deceptively conclusive output.

The University of California, Berkeley, admissions sex-bias case perhaps best explains Simpson's paradox (Bickel, Hammel, and O'Connell, 1975). In this case, a group of women filed suit based on an overall university acceptance rate that seemed to show a distinct bias against women: Men were accepted at a much higher rate. However, when the data were examined in greater detail, the acceptance rate for each specific school within the university showed that *women* generally had an advantage. The ramifications for such a paradox in force-ratio modeling are multifarious. For example, in a specific case, a force ratio might be wholly insufficient in 90 percent of the country but adequate in 10 percent. While the overall force ratio would presumably be very low, the counterinsurgents might still succeed if the adequately covered areas were also the most strategically vital. This would not be uncommon in cases in which governments focused resources on protecting vital infrastructure or population centers. Conversely, the government might cover 90 percent of the AO adequately but miss the critical 10 percent and lose. In these cases, both the specific and aggregated data would be deceptive and undermine the validity of the case. Simpson's paradox also applies when cases are examined as part of a data set, potentially undermining the validity of an entire study.

We clearly state in the summary of *How Insurgencies End* that our findings should not be taken prescriptively. Despite this warning, and the sobering caveats found in Quinlivan's various articles, mathematical ratios can appear—even unintentionally—as prediction. Policy advisers should take great care when discussing or proposing specific COIN force ratios based on historical data.

If, as we suggest, statistical force ratios are unreliable, then, in preparation for prospective campaigns, political leaders and strategic planners will need to depend on exacting staff and academic work to acquire the necessary data to determine force levels; the best analysis of this type is detailed, often down to the subdistrict level. Acquiring these data, especially in denied areas or failed states, will strain both military and civilian collection resources. Staffs will have to pore through volumes of quantitative and qualitative reporting, judging the content, sources, and authors of the reports in order to achieve the kind of fidelity Maj. Gen. Michael T. Flynn, the senior military intelligence officer in Afghanistan in early 2010, describes in his critique of current intelligence practices in COIN (Flynn, Pottinger, and Batchelor, 2010). These are the kinds of squishy, single-case inputs that are anathema to case-study modeling but also are the most effective tools for gauging success and determining requirements in COIN operations.

During operations, policymakers will have to depend on the plain-word assessments of their trusted field commanders in order to conduct force planning and to shape insurgency endings. If they do not trust their commanders to give them accurate assessments, it is their responsibility to replace them with more-competent officers. And, at each stage of analysis, military and civilian planners will have to fight the urge to streamline reports or to translate qualitative data into graphical, quantitative presentations or one-page executive summaries that do not capture the complexities of the problem.

For counterinsurgents faced with the inescapable reality of limited resources, especially in the latter stages of an unpopular campaign, civil-defense forces (CDFs) or militias may be used to offset the need for more uniformed troops and police. The next section addresses this option.[9]

---

[9] It may be possible to develop alternative metrics to determine force ratios, but proposing and discussing those methods falls outside the scope of this monograph. We note that, in nearly every successful COIN within our data set, an increase in troop levels proved nec-

Civil-Defense Forces

We initially tasked our researchers with examining CDFs in concert with barrier plans and the relative success counterinsurgents had in turning insurgents; this finding is included in Appendix B. In light of recent events in both Iraq and Afghanistan, it is apparent that CDFs should have been addressed as a separate variable. Counterinsurgents commonly employ some variation of civil-defense programs. These programs—whether termed *militia, home guard,* or *self-defense forces*—figured prominently in some of the most successful and oft-cited COIN operations. In these campaigns, a CDF helped establish physical security by supplementing government forces while also fostering a perception of security and connectivity with the government. McCuen (1966, p. 107) places high value on CDFs:

> An important part—possibly the most important part—of counter-organization of the population is the organization of its self-defense against revolutionary intimidation and exactions. Unless the people themselves have the means and commitment to resist, their desire for personal security is likely to overpower their loyalty to the government or neutrality. Even in [the] early phase of the war, organization of local auxiliary police and militia units should be a first priority task of the governing authorities.

O'Neill (1990, p. 130) points out that not only are civil militias a viable COIN tactic but they can serve to buffer fledgling local governments:

> In order to free regular military forces for counter-guerrilla operations and to provide security for government officials . . . local self-defense forces may be established. Where they are not, civilian officials who are in charge of social and economic programs

---

essary to defeat an active (vice proto) insurgency: Greece, Peru, Angola, Malaya, Kenya, Algeria, possibly Colombia, and Iraq, among others. An increase in troops often leads to an increase in security, which we have identified as a fundamental necessity in COIN. We found the quality and behavior of these troops to be equally important to their numbers, if not more so.

can be intimidated or eliminated by insurgent violence; witness the plight of unprotected mayors in El Salvador in 1988 who either resigned or were assassinated.

He goes on to point out that implementing a CDF program can be tricky. To be effective, it should complement local governance programs but must also be perceived as a legitimate alternative to insurgent presence:

> When local militias are established to prevent [insurgent intimidation], their effectiveness will be partly contingent on whether they constitute a disciplined force perceived to be a servant of the people, as in the case of the *firqats* in Oman, or are instead ill-disciplined units guilty of excesses against the people, as in the case of the Civilian Home Defense Force in the Philippines. (O'Neill, 1990, p. 130)

Our selected experts generally concurred that CDFs are a useful tool, both for the counterinsurgent and for the insurgent. Mao and Giáp exhort communist insurgents to make use of militias to serve not only as self-defense forces but also as a basis for the progressive development of the insurgent cadre into a conventional force. This mirrors COIN theory, which envisions an eventual absorption of CDFs into the constabulary or the national army. Mao specifically describes seven types of guerrilla organizations, one of which is a local militia. He believed that militia "should be formed in every locality" (Mao, 1961 [2000], p. 75). Here, he describes the communist Chinese strategy against the occupying Japanese forces during World War II:

> All the people of both sexes from the ages of sixteen to forty-five must be organized into anti-Japanese self-defense units, the basis of which is voluntary service. As a first step, they must procure arms, then they must be given both military and political training. Their responsibilities are: local sentry duties, securing information of the enemy, arresting traitors, and preventing the dissemination of enemy propaganda. . . . Such units are reservoirs of manpower for the orthodox forces. (Mao, 1961 [2000], p. 80)

Giáp, who often parrots or closely interprets Mao's ideology and teachings, envisions a similar formula:

> Our Party advocated that, to launch the people's war, it was necessary to have three kinds of armed forces. It attached great importance to the building and development of self-defense units and guerrilla units. Militia was set up everywhere. (Võ, 1961 [2000], p. 141)

For the counterinsurgent (and perhaps the insurgent), proper timing is necessary not only to ensure that CDFs survive through inevitable early-stage vulnerabilities but also to ensure that the program enjoys the kind of grassroots support necessary to build and sustain both a local and national program. Intent to establish such programs is irrelevant if the population is unwilling to participate. In some of the cases explored in this study, the counterinsurgents tried to establish programs while the population remained thoroughly unconvinced that it was in its best interest to side with the government. The failure of the Iraqi Civil Defense Corps is only one case in point. McCuen (1966, pp. 110–111) points out that it is more difficult to build a CDF in the midst of an ongoing COIN campaign. Here, he describes the French experience in Indochina:

> They obviously did not have the opportunity to form either auxiliary police or militia before the fighting started. . . . As the British found in Malaya, screening recruits after the fighting starts is most difficult. As a result, the Vietminh deeply infiltrated much of the militia. They made a special effort to attack indigenous units—particularly the good ones. Without nearby French troops to back them, few Vietnamese were interested in being members of the village self-defense forces. . . . The villagers' lack of training and equipment made them no match for the tough, battle-hardened guerrillas. . . . Of course, the French could not keep the militia effective under these conditions.

McCuen also points out one of the inherent dangers of building a CDF that may or may not have strict loyalties to the central government. Central governments may have legitimate concerns regarding the capabilities and intent of armed militias, especially when those militias signify a lack of government security capacity. Here, McCuen (1966, p. 228) describes the French efforts to develop CDFs under Special Administrative Sections (SASs)—essentially district offices—across the country. Timing is a common theme:

> The first thing that the S.A.S. had to do was to establish its security and that of the village. . . . [They] organized a village self-defense unit as soon as possible. Such a unit might consist of 100 local Muslims armed with a wide variety of weapons. The problem of the S.A.S. was to choose the correct moment when the population was sufficiently won over that it could distribute rifles without danger of having them passed to the rebels.

Many concerns must be addressed if the counterinsurgent is to develop and sustain a viable program. Planners must decide whether the CDF members will be drawn from a pool of volunteers or receive pay, whether they will be supplemented with regular army cadres or simply provided equipment and rudimentary training, and whether they will serve part or full time. In each of the cases we examined, the counterinsurgents shaped their CDF program to fit the unique circumstances of their environments, sometimes successfully, sometimes not. McCuen (1966, p. 108) recognizes the need for case-specific flexibility but also offers generalized best practices:

> Supported by the regular military forces dispersed throughout the country, these militia units should be hand-picked from among the people, staffed with local reservists and, if possible, commanded by small cadres of regulars. The governing authorities should make membership in the militia desirable. They can do this both by paying for duty periods and by social activities and various privileges. . . . In this manner, important members of the local community can be committed to the government. At the same time, the militia affords the population and its leaders a means of resisting revolutionary intimidation.

While the French were unsuccessful in leveraging CDFs in what was then Indochina, the Marine Corps saw considerable, if short-lived, success with Combined Action Program (CAP) in South Vietnam. Robert M. Cassidy (2008, p. 138) describes how the Marines shaped CAP to fit local circumstances:

> The CAP was a local innovation with potentially strategic impact—it coupled a Marine rifle squad with a platoon of local indigenous forces and positioned this combined action platoon in the village of those local forces. . . . The mission of the CAP was to destroy the [Vietcong] within the village or hamlet area of responsibility; protect public security and help maintain law and order; protect friendly infrastructure; protect bases and communications within the villages and hamlets; organize indigenous intelligence nets; participate in civic action; and conduct propaganda against the Viet Cong. . . . In this way, a modest investment of U.S. forces at the village level or local level can yield major improvements in local security and intelligence.

An Office of Naval Research (ONR) study determined that CAP marines accounted for 7.6 percent of (Marine-related) reported enemy killed in action (KIA) while suffering only 3.2 percent of Marine casualties over a four-year period between 1965 and 1969 (Allnutt, 1969). While these statistics may not tell a complete story, ONR also quotes two South Vietnamese officers on CAP. The first quote is from the sector chief for regional force/popular force troops, Quang Tri province, while the second quote is from the commander of all popular force troops in I Corps, South Vietnam. One must assume that both officers were somewhat biased due to their close association with the program, although their thoughts are echoed by other reports:

> I would emphasize that in thinking about CAP teams, we must view them from both a military and political point of view. The important thing politically is that the CAP team symbolizes American presence in Viet Nam. By their behavior, the CAPs refute VC [Vietcong] propaganda. They show the people that the U.S. presence is different than that of the French. (Allnutt, 1969, p. 11)

What can one company of regular troops do, operating in an area? Compare this with ten CAPs—going on patrols, setting ambushes, doing some civic action—they're really having an impact on 30,000 people. I'd pick one Combined Action Company over a battalion of infantry, if I had a choice. We need some big units, yes, but in general this war is for the people. (Allnutt, 1969, p. 12)

The British developed a similar organization in Malaya called the Home Guard. British officers specifically designed this CDF to secure villages that had been transplanted under the strategic-hamlet program. Home Guard members served as volunteers and, in most cases, served only to provide night watch and limited patrolling. Some reporting puts the Home Guard end strength at approximately 250,000. Whether or not this statistic is accurate, the British credit the Home Guard program with helping to successfully end the insurgency (McCuen, 1966, p. 161). British advisers also crafted a successful CDF program in Oman during the campaign against the Dhofari insurgency. Omani *firqat* (Arabic plural for team or unit) teamed with civil-action teams to provide a mutually reinforcing civil-affairs effort:

[The] key role in civil affairs from 1971 onwards was played by the firqat forces, supplemented by the Civil Action Teams (CAT). . . . Engineers would drill wells, and build a shop, school, clinic and mosque. Dhofaris would cluster around these ad hoc settlements for food, water, medical and veterinary care. . . . Civilians would in turn provide both intelligence and volunteers for the government's tribal militias. The CAT program combined "hearts and minds" work among Dhofaris with the tactical requirement of population control. It not only provided tangible evidence of the new Sultan's commitment to the welfare of his subjects, but it also helped concentrate the civil populace [and] encouraged desertions from the guerrillas. . . . (G. Hughes, 2009, p. 290)

Both the Greeks (1945–1949) and Peruvians (1981–1992) had success in building volunteer militias that played key roles in bringing about successful ends to their respective insurgencies.[10] Greece put reserve officers in charge of the National Guard Defense Battalions (Tagmata Ethnofylakha Amynhs, or TEA), giving them an opportunity both to serve in their home regions and to free up regular officers for other duties:

> The men serve on a part-time basis while maintaining their civilian occupations as farmers, shopkeepers, clerks, mechanics, and so forth. They receive no pay; however, T.E.A. members do get priority in the occasional distribution of aid, international gifts, boots, and other equipment. Apparently, their main motivation is the realization that it is necessary to protect their homes against the rebels who so terrorized them during the war. The T.E.A. are very lightly equipped, having only such weapons as rifles, submachine guns, and light machine guns. . . . Training is mainly conducted on Sundays and holidays. . . . A significant element of the T.E.A. training is political. . . . The men rotate guard duty at night. They actively patrol, especially along the frontier. (McCuen, 1966, p. 112)

The Peruvian *rondas*, or CDFs, were both a by-product of Shining Path atrocities and the result of deliberate Peruvian military action. The *rondas* were successful probably because individual members were strongly motivated to defend their properties and kin against the insurgents:

> Shining Path reacted [to the creation of CDFs] by increasing violence against the peasantry. But all this achieved was the proliferation of *rondas*, or "Committees of Civil Auto-Defense," to the point that, by 1990, Sendero [Luminoso] had become trapped in a kind of trench warfare against the peasants. This constituted the first strategic victory for the Armed Forces and the first real defeat of the Shining Path since the war had started. . . . [By] recruiting youth who were allowed to do their obligatory military service in their own communities, and by distributing weapons

---

[10] Temporarily for the Peruvians, who would see Shining Path return a decade after the capture of Guzmán.

to the *rondas*—even though these arms were merely shotguns—the Armed Forces, and the state they represented, demonstrated that they had obtained hegemony in the zone. (Degregori, 1998, pp. 146–147)

CDFs are often credited with providing excellent local intelligence to police and military units, gathering information that would otherwise be unobtainable through more-formal intelligence-collection means (McCuen, 1966, p. 112, referring to the Greek case).[11] When counterinsurgents implement a successful CDF program, police and police intelligence (special branch) officers leverage their relationships with local leaders and militia commanders to build and expand their human-intelligence networks. CDFs generally proved to be both a force multiplier (in that they gave local advantage to the counterinsurgents over the insurgents) and an economy-of-force measure (in that they amplified counterinsurgent force levels, expanding the reach of the government in the absence of sufficient regular security forces). In some cases, the CDFs were successful in encouraging defections, as locally recruited insurgents observed stabilization and, eventually, some level of normalization in their home territories.

Qualitative analysis of our data set seems to show that volunteer (unpaid), part-time forces are generally more effective and longer lasting than paid, full-time forces, but this is not always true. While unpaid part-timers are often conscientious in their duties, they rarely succeed without the motivation of indiscriminate insurgent terror or locally applied government reform programs. Paid full-time militias can be thrown together quickly, giving the counterinsurgent a chance to establish immediate security in a specific area while taking young males off the streets. However, since the paid militia members are primarily motivated by money, they also are more vulnerable to infiltration, bribery, desertion, and defection. We cannot prescribe one or the other for any specific condition.

---

[11] Geraint Hughes (2009) provides conflicting evidence of CDF intelligence reporting but seems to find some value in *firqat* reporting.

Incorporating a CDF program into a COIN campaign plan can help shape a positive ending, but only if the program is well designed and systematically implemented. Security-force commanders tasked with building a CDF have to display flexibility to meet variations in local conditions. For example, one village may be motivated to build a CDF because of a tribal dispute, while another may be enticed by government protection or largesse. However the program is implemented, it can be gauged a success if it helps establish security and demonstrates a shift in popular perception toward the government. McCuen (1966, p. 112) views the Greek TEA as a model CDF program:

> Faced with such armed, trained and determined nucleuses among the people, the revolutionaries know that they can no longer recruit and terrorize as they once did. They can no longer hide in the villages without being detected and reported by their neighbors. The Greek rural populations can now defend themselves.

Soviet and Afghan communist efforts to build CDFs in southern Afghanistan in the late 1980s and early 1990s were successful in that they helped secure the south for several years after the Soviet withdrawal. However, the eventual collapse of the Afghan government and the southern militias in the face of the Taliban also highlights weaknesses in Soviet COIN theory. As part of their increasingly savvy COIN strategy, the Soviets established CDFs in Helmand and other southern provinces in order to offset mujahideen presence and the lack of sufficient Soviet army forces. While Helmand lacked a homogenous tribal structure like some of the other provinces, the Soviets took a top-down rather than grassroots approach to the problem. They demanded strict screening for political and religious ideology and brought in leaders from Kabul, alienating the locals. They paid militia but failed to closely monitor the CDFs, allowing warlordism to flourish. Afghan scholar Antonio Giustozzi (2008) saw the Soviet CDF program in the south—a program put into practice as the Soviets were preparing to leave—as a step toward feudalism.

The success of the current coalition CDF programs in Iraq and Afghanistan had yet to be determined by late 2009. These programs offer well-documented and contrasting models. The Sons of Iraq (SoI) program recruited tens of thousands of young Sunni into a national, full-time, paid militia. SoI is credited with "draining the swamp" in some of the more violent provinces in Iraq. Coalition forces in Afghanistan, meanwhile, have taken several different approaches to building CDFs. The Afghan Public Protection Program (AP3) is a national, paid force while regional variations of *arbakai*, or traditional tribal/ethnic militia, are unpaid volunteer forces beholden to local councils. Each model has its virtues and inherent dangers, and each is necessarily tailored to fit regional and (in the case of the *arbakai*) local conditions. A close examination of each program is warranted in the comprehensive histories that eventually will be written on Operations Enduring Freedom and Iraqi Freedom.

CHAPTER SIX

Conclusions

Our conclusions reflect both the intersection and dichotomy between quantitative and qualitative approaches to case-study research. While it would not have been possible to draw generalized conclusions about insurgency endings without a close examination of a sizable data set, the lack of control over the data necessitates educated interpretation to a degree that might bring discomfort to those familiar with strict scientific examination. This middle-of-the-road approach prevents us from offering conclusive or predictive findings: None of our quantitative analysis stands alone, while our broader analysis stands as a singular interpretation of the history of modern insurgency endings. Further, we recognize that our quantitative study failed to adequately address some critical elements of COIN, including, but not limited to, information operations, criminalization, force ratios, and CDFs. Our research should be compared and contrasted with other, similar studies, several of which we have cited herein.

With these final caveats in place, there are some generalized lessons on which counterinsurgents might draw when shaping individual campaigns. Each should be examined and, if found applicable, modified to fit specific conditions. So while a counterinsurgent should not look at the strong correlation between a loss of insurgent sanctuary and government victory and then put all efforts toward interdicting sanctuaries, it might be prudent to incorporate some form of interdiction operations into a comprehensive campaign plan when sanctuary is present. Some, if not most, of our findings are best considered during the planning stages of a COIN operation, but all should be reconsidered during periodic shifts in campaign emphasis and direction.

While we will not reiterate each quantitative finding here, we note that governments seem to have some advantages over insurgents in the majority of cases. At least in the early stage of a campaign, the government is typically better organized, has a stronger military capability, and has leaders or bureaucrats experienced in overcoming at least basic systemic challenges. Governments seem to fare better without external support, while insurgents almost always depend on sponsorship. Government military organizations typically operate out of established bases, while insurgents struggle to locate and maintain safe haven. Government repression often is effective in tamping down, even if temporarily, insurgent violence, while the use of indiscriminate insurgent terror correlates with insurgent defeat. Populations increasingly live in large urban areas where the government has a number of advantages. How then, do insurgents win?

Based on both our quantitative and qualitative findings, it seems that governments defeat themselves more often than they are defeated by a dominant insurgency:

- Governments ignore the insurgency until it develops into a credible threat.
- Governments fail to address root causes.
- Governments address root causes half-heartedly or too late, stoking discontent.
- Governments fail to identify major shifts in strategic momentum.
- Governments fail to extend credible control into rural areas.
- Governments become dependent on a fickle sponsor.

External sponsors of governments embroiled in COIN campaigns also fail to pay attention to indigenous discontent, intelligence indicators, and rural sanctuaries. With the exception of the Soviet Union and China in the 1960s and 1970s, they find it especially difficult to sustain operations over time in the face of changing domestic priorities and faltering political willpower. Our research reinforces the widely held belief that democratic sponsors of external campaigns are especially vulnerable to shifts in domestic priorities. Gil Merom (2003, p. 229) states that, "After 1945, democracies discovered that military superiority and battlefield advantage have become fruitless, if not counterproductive, in protracted counterinsurgency campaigns." Three of our selected experts—Merom, Weinstein, and Kalyvas—describe in detail how democracies and other governments give way to insurgencies under the weight of their own corruption, weakness, incompetence, or the general failure to address simple tenets of security and good governance.

In the cases in which they are not defeated by a competent government, insurgents sometimes defeat themselves. In many of these cases, they fall victim to their own violent tendencies or near-pathological need to invest the populace with strict dogma (e.g., Peru, Guatemala). When the military balance seemed to tip in their favor, some insurgents became overly ambitious and perhaps arrogant, leaving themselves vulnerable to a conventional military defeat (e.g., Greece, Sri Lanka). Many of our mixed cases showed that insurgents avoided defeat only by acceding to demands they might have shunned under more-favorable circumstances. While, by definition, they also benefited from the final settlement, a closer examination of some cases might show that they were co-opted.

No matter how one applies individual quantitative lessons, favorably ending an insurgency remains a matter of conducting a well-timed, aggressive, fully resourced, population-centric campaign designed to address the root causes of the conflict. Our research found no "COIN shortcuts" or generalized exceptions to these onerous obligations: Half-measures and inconsistent application of basic COIN tenets presage defeat or ongoing violence. External sponsors are further burdened with the need to sustain domestic support over an extended period of time. In this way, insurgency is no different from any other form of warfare: It is a conflict of opposing wills. If the insurgents can shift the opinion of the sponsor's population against the war (e.g., Vietnam), they then stand a good chance of forcing a withdrawal and ending the war in their favor. If the sponsors set realistic expectations at home and devote sufficient resources to the fight, they stand a good chance of shrugging off insurgent violence and propaganda long enough to win:

When sponsoring external interventions, at the strategic level, "gaining and maintaining U.S. public support for a protracted deployment is critical" (FM 3-24, p. 1-24).

These last conclusions are intended to inform COIN campaign planning and midstride campaign adjustments.

With a few exceptions, lasting insurgency endings are shaped not by military action but by social, economic, and political change. At their core, insurgencies are battles for the control of public support. Therefore, violence is useful to the insurgents only when it creates a level of instability that allows them to effect social change in their favor, and violence is useful to the government only when it helps stabilize the population long enough to effect lasting change in its favor. The government may defeat the insurgent military cadre, but, with few exceptions, insurgencies do not end until case-specific root causes are addressed: The kind of grassroots support necessary to build and sustain an insurgency is fed on social, economic, and political discontent.[1] If a government successfully addresses root causes, it is possible to defeat an insurgency without defeating the insurgents themselves. Conversely, as we saw in Vietnam, Algeria, and South Africa, it is possible to lose while defeating or suppressing the insurgent cadre.

One could argue that, since insurgencies stem from root-cause discontent and *lasting* victory necessitates the government address the root causes of the insurgency, insurgents—or at least the people they represent—win every clearly decided case. This could be true even in cases in which the insurgent cadre is wholly annihilated but might not necessarily apply to insurgencies or terrorist groups that never represented a legitimate social movement or cause.

Government victories often cause the insurgency to splinter, leaving behind small elements of irredeemables that may or may not represent an ongoing threat. Tracking these splinter groups can provide tremendous insight into the nature of the insurgency ending. We noted that, when the government is winning and the insurgency is in its "tail" phase, often a small group of insurgents splinters away from the cadre or leader-

---

[1]   The October 2009 joint publication (JP) on counterinsurgency, JP 3-24 (p. xii), uses the term *core grievances* to describe root causes of insurgency. Either term is applicable.

ship group.[2] In some cases, this splinter element is an irredeemable fringe unwilling to negotiate or enter into an amnesty program. In others, the splinter is formed from an irredeemable core leadership element. In both cases, the splinter groups are intent on continuing the struggle against the government in one form or another. They may try to sustain or reignite the insurgency immediately, or they may be willing to hibernate until they see an opportunity to reemerge. Counterinsurgents should be able to tell a great deal about their long-term prospects by observing how the general populace perceives this splinter group. If the irredeemables are absorbed and protected by the indigenous population and are able to demonstrate continuing grassroots support, then the insurgency probably has not ended. If, however, an exhausted or government-leaning populace ostracizes these irredeemables, they become little more than an isolated terrorist cell. It is at this point, when the root causes of the insurgency have been addressed and all willing insurgents reconciled, that killing might end the insurgency.

Professional intelligence organizations should be able to identify shifts in strategic momentum during the course of a campaign by incorporating a small set of generalizable indicators into the all-source analysis process. All forms of warfare require organized militaries to conduct thorough all-source intelligence collection and analysis. With exception, no one indicator is sufficient to indicate a major shift in strategic momentum over the course of a lengthy campaign. This may be especially true in complex COIN environments with no definable front line and an enemy that may or may not be in uniform and that may or may not possess identifiable bases or equipment. COIN requires the development of a unique and, we argue, locally adapted set of intelligence metrics. As long as all caveats are considered, tracking defections, desertions, and the flow of voluntarily provided human intelligence could help intelligence professionals identify shifts in momentum from the tactical to the strategic levels of war. At best, these indicators can help identify a tipping point along the timeline arc of the campaign. The

---

[2] Several noted terrorism experts, including Walter Laqueur (1976), have ably described the splintering of terrorist organizations. Similar dynamics apply to insurgency leadership in the final stages of a losing effort.

value of defections also points to the value of reconciliation as part of a COIN strategy.

No insurgency ending is inevitable. It is a simple thing to say, "Nothing is inevitable." It is quite another to turn a COIN campaign once it has reached a tipping point. Many of the cases we studied, however, showed that a renewed burst of purposeful activity (e.g., Sri Lanka's military campaign against the LTTE) or unexpected happenstance (e.g., the coup d'état in Portugal that led to the collapse of the colonial government in Angola) could force a major shift in momentum at any point along the conflict arc. When coupled with the finding that insurgents do not win simply by sustaining operations over time, the prospects for counterinsurgents appear more hopeful. Counterinsurgents should, however, be aware that these renewed bursts of activity typically precede lengthy and expensive commitments to end the insurgency.

Further, as David Kilcullen and other experts make clear, while it is necessary to apply basic COIN principles, there are no cookie-cutter approaches to COIN. Counterinsurgents must adapt tactics, operations, and strategy to fit specific circumstances and then be prepared to change them as often as necessary to win.

Case Studies: Methodology

Selection Process

After establishing the categories of insurgency endings, the next step was to create a database of insurgencies for the quantitative approach. The data set of insurgencies was drawn from James Fearon and David Laitin (2003a).[1] The 127 insurgencies Fearon and Laitin examined were ones that met the following three criteria:

1. They involved fighting between agents of (or claimants to) a state and organized, nonstate groups that sought to take control of a government, take power in a region, or use violence to change government policies.
2. The conflict killed or has killed at least 1,000 people over its course, with a yearly average of at least 100.
3. At least 100 people were killed on both sides (including civilians attacked by rebels).

A distinction should be made between conflicts whose aims are those of an insurgency and those conducted as a classic (e.g., Maoist) insurgency might be. For instance, some of the insurgencies analyzed here (e.g., the 1970 conflict between the Palestine Liberation Organization [PLO] and Jordan) have featured operations more characteristic of conventional than irregular conflict. Even wars whose aims could

---

[1] A number of other excellent data sets are available, including those developed and coded by the Dupuy Institute, the Center for Army Analysis, and Jason Lyall and Isaiah Wilson III.

have been pursued through unconventional means (e.g., the attempted secession of Biafra from Nigeria) were actually fought more like conventional wars, with armies directly confronting one another. Nevertheless, we took a broad view of insurgency and did not automatically reject conflicts from consideration because their operations did not assume classic form.

We then added and subtracted conflicts from this list of 127. Because the Fearon-Laitin data end in 1999, we wanted to add conflicts that crossed the 1,000-death threshold after 1999; there were 11.[2] We also added two conflicts (Namibia's independence movement and the Tupamaros in Uruguay) that merited reexamination. In the Philippines, we coded two conflicts where the Fearon-Laitin list had just one; in Tibet, the reverse. We then excluded conflicts that, in our opinion, had few if any useful similarities to the current insurgencies that motivated our research interest. Excluded cases tended to be more like countercoups and insurrections; the 51 conflicts excluded are listed in Appendix D.

The 89 insurgencies included in this analysis are listed in Table A.1, sorted by start date. Each insurgency is labeled by country (or region), together with additional descriptors in the case of potential ambiguity. Following the name is the start and the end dates (as we have calculated them—the Fearon-Laitin calculations differ in certain cases).

The basic approach taken in the quantitative assessment was to group all the insurgencies by how they fit each possible value for it (e.g., how many insurgencies sought independence, how many wanted a Marxist government). For the total number of insurgencies associated with each value, we then counted how many of these insurgencies resulted in which of four outcomes—that is, how many insurgencies in this category were won by government, lost by government, had mixed outcomes, or are still ongoing (or were as of this writing).[3]

---

[2] They were India's Naxalite insurgency, Uganda's Arab Deterrent Force (ADF), Kosovo, Israel's Intifada II, Iraq, Afghanistan after the Taliban won and again after they fell, the Niger delta, Darfur, the Ivory Coast, and southern Thailand.

[3] For conflict duration, we used the converse methodology: Insurgencies were divided into four outcome categories, and then the relevant characteristics of each outcome category were assessed.

Table A.1
Insurgencies Examined for This Study

Insurgency	Government Won	Insurgents Won	Mixed	Ongoing
	Outcome			
China, 1934–1950		x		
Greece, 1945–1949	x			
Philippines (Huk), 1946–1955	x			
Indochina, 1946–1954		x		
Burma, 1948–2006	x			
Malaya, 1948–1960	x			
Colombia (La Violencia), 1948–1962			x	
Kenya, 1952–1956	x			
Cuba, 1953–1959		x		
Algerian independence, 1954–1962		x		
Lebanon, 1958–1959	x			
Indonesia Darul Islam, 1958–1960	x			
Tibet, 1959–1974	x			
Congo/Katanga, 1960–1965	x			
Guatemala, 1960–1996	x			
South Africa, 1960–1994		x		
Namibia, 1960–1989		x		
Eritrea, 1960–1993		x		
Laos, 1960–1975		x		
South Vietnam, 1960–1975		x		
Iraqi Kurdistan, 1961–1974	x			
Mozambique, 1962–1974		x		
Guinea-Bissau, 1962–1974		x		

Table A.1—Continued

Insurgency	Outcome			
	Government Won	Insurgents Won	Mixed	Ongoing
Angolan independence, 1962–1974		x		
Yemen, 1962–1970			x	
Uruguay, 1963–1973	x			
Colombia (FARC), 1963–				x
Zimbabwe, 1965–1980		x		
Dominican Republic, 1965–1966			x	
Biafran secession, 1967–1970	x			
Argentina, 1968–1979	x			
Cambodia, 1968–1975		x		
Northern Ireland, 1969–1999	x			
Philippines (NPA), 1969–				x
Jordan, 1970–1971	x			
Philippines (MNLF), 1971–1996	x			
Bangladesh, 1971–1972		x		
Baluchistan, 1973–1977	x			
Angola (UNITA), 1975–2002	x			
Morocco, 1975–1991	x			
East Timor, 1975–2000			x	
Lebanese civil war, 1975–1990			x	
India northeast, 1975–				x
Indonesia (Aceh), 1976–2005	x			
Mozambique (RENAMO), 1976–1995			x	
Sri Lanka, 1976–				x

Table A.1—Continued

Insurgency	Outcome			
	Government Won	Insurgents Won	Mixed	Ongoing
Philippines (MILF), 1977–2006	x			
Nicaragua (Somoza), 1978–1979		x		
Afghanistan (anti-Soviet), 1978–1992		x		
Kampuchea, 1978–1992			x	
El Salvador, 1979–1992			x	
Somalia, 1980–1991		x		
Senegal, 1980–2002			x	
India (Naxalite), 1980–				x
Peru, 1981–1992	x			
Nicaragua (Contras), 1981–1990			x	
Turkey (PKK), 1984–1999	x			
Sudan (SPLA), 1984–2004		x		
Uganda (ADF), 1986–2000	x			
Uganda (LRA), 1987–				x
Papua New Guinea, 1988–1998			x	
Liberia, 1989–1997		x		
Kashmir, 1989–				x
Rwanda, 1990–1994		x		
Moldova, 1990–1992		x		
Sierra Leone, 1991–2002	x			
Somalia (post-Barre), 1991–				x
Nigeria (Niger Delta), 1991–				x
Algeria (GIA), 1992–2004	x			
Croatia, 1992–1995	x			

Table A.1—Continued

Insurgency	Government Won	Insurgents Won	Mixed	Ongoing
Afghanistan (post-Soviet), 1992–1996		x		
Tajikistan, 1992–1997			x	
Georgia/Abkhazia, 1992–1994			x	
Nagorno-Karabakh, 1992–1994			x	
Bosnia, 1992–1995			x	
Burundi, 1993–2003			x	
Chechnya I, 1994–1996			x	
Afghanistan (Taliban), 1996–2001		x		
Zaire (anti-Mobutu), 1996–1997		x		
Kosovo, 1996–1999		x		
Nepal, 1997–2006			x	
Congo (anti-Kabila), 1998–2003			x	
Chechnya II, 1999–				x
Israel, 2000–				x
Afghanistan (anticoalition), 2001–				x
Ivory Coast, 2002–				x
Darfur, 2003–				x
Iraq, 2003–				x
South Thailand, 2004–				x

NOTE: Huk = Hukbalahap. NPA = New People's Army. RENAMO = Resistência Nacional Moçambicana, or Mozambican National Resistance. MILF = Moro Islamic Liberation Front. SPLA = Sudan People's Liberation Army. LRA = Lord's Resistance Army.

Characterizing Results

We characterize our results in a number of ways. We begin with a discussion of three sets of factors that apply to all insurgencies: general, income and urbanization, and terrain. We then move to a discussion of factors that pertain to insurgents and follow that with a discussion of those that pertain to the government. We then describe what the multivariate regression analysis of our data tells us, and we follow that with a discussion of ways to conclude insurgencies. Finally, we offer some conclusions about what the various factors tell us about ending insurgencies.

We limited ourselves to insurgencies that passed a certain threshold; however, many protoinsurgencies die before reaching this threshold. Thus, while concluding, for instance, that most insurgencies that fought for independence succeeded, we omit all the protoinsurgencies that sought their country's independence but never achieved sufficient momentum to make the threshold for inclusion. This may constitute a form of sample bias that skews results.

Every insurgency was assigned to one of the four outcomes based on our assessment. In 28 cases, the government won. In 26 cases, we judge the insurgents to have prevailed. In 19 cases, we view the outcome as mixed, in that neither side achieved all it wanted. Sixteen insurgencies have yet to conclude.

We then parceled out the 89 insurgencies to various RAND analysts and research assistants, most of whom had enough knowledge of the region or insurgency to reach conclusions on their character based on earlier research. This exercise was performed twice. The first survey was carried out in the spring of 2006 with an emphasis on the factors that led to the government's winning and losing. The second survey was carried out in the autumn of 2006 with more of an emphasis on how the insurgencies were brought to an end.

Efforts were made to ensure that the answers were consistently coded. In cases in which they were not, adequate translation was made—e.g., what one analyst put down as "yes," "somewhat," and "no" were converted into "high," "medium," and "low." The original database retains some of the finer distinctions made by analysts (e.g., between medium-high, medium, and medium-low), but, in many cases, these finer distinctions were suppressed to facilitate statistical analysis (but often retained in the database as spreadsheet comments). Finally, we tried hard to generate a variable for all cases for which one was relevant or meaningful; in practice, this meant that several indicators that were not well known were parameterized toward the middle of the option space.

Supplemental Findings

Location

Insurgencies have generally been a phenomenon of the developing world (Northern Ireland aside) (see Figure B.1).[1] Generally speaking, governments tend to do roughly as well against insurgents across every area of the world, with one conspicuous exception: sub-Saharan Africa (see Table B.1). Part of the African difference lies in the large number of insurgencies that come under the rubric of decolonization (Guinea-Bissau, Angola, Mozambique, and Kenya), or, what is similar, the struggle for black majority rule (Zimbabwe, Namibia, and South Africa). If those seven (one government victory and six government defeats) are excluded, the record in that region is 4:4:7 (four government victories, four mixed outcomes, and seven government defeats) rather than 5:4:13—which still makes the region an outlier but not dramatically so.

Several countries may find themselves overrepresented in any insurgency list. Six insurgencies were fomented (or put in motion) by the breakup of the Soviet Union (Moldova, Georgia/Abkhazia, Tajikistan, Nagorno-Karabakh, Chechnya I and Chechnya II)[2] and three by the breakup of Yugoslavia (Croatia, Bosnia, and Kosovo). Vietnam was

---

[1] Spain's Euskadi Ta Askatasuna (ETA, or Basque Homeland and Freedom) missed the cutoff in terms of total casualties (and they were never generated at a rate of 100 per year required to meet another of the Fearon-Laitin inclusion requirements).

[2] Tajikistan and both Chechnya insurgencies were classified as Middle East because of their radical Islamic component.

Figure B.1
Location of Insurgency

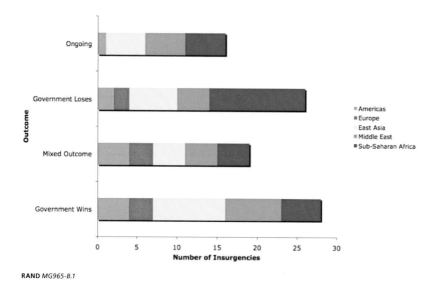

RAND *MG965-B.1*

Table B.1
Outcomes as a Function of Where the Insurgency Took Place

Outcome	Americas	Europe	East Asia	Middle East	Sub-Saharan Africa
Government wins	4	3	9	7	5
Mixed outcome	4	3	4	4	4
Government loses	2	2	6	4	12
Ongoing	1		5	5	5

directly or indirectly associated with five of them: Indochina, South Vietnam, Laos, Cambodia (1968–1975), and Kampuchea (1978–1992). India faces three ongoing insurgencies (Kashmir, Naxalite, and the northeast). The Philippines has faced four, one of which it repulsed outright (Huk), two in which it is the putative winner (the MILF and MNLF) and one of which is dragging on (NPA). Indonesia has seen three insurgencies (Darul Islam, Aceh, and East Timor). Many countries have experienced two.

Regional Religion

Given today's news, it was deemed useful to see whether the pattern of insurgency outcomes within the Muslim world differed from that outside it. The data suggest that governments completely within the Islamic world are somewhat less apt to lose and far less apt to settle for a mixed outcome than countries completely outside the Islamic world (see Figure B.2 and Table B.2). When a decision was reached, governments there won more than half of the time, while governments elsewhere won only a third of the time. We take note, however, of how many insurgencies are separatist movements by Islamic regions from countries that are not Islamic, or separatist movements by non-Islamic regions from countries that are Islamic. Indeed, the last two categories (which account for just over 10 percent of the total insurgency database) account for five of the 15 ongoing insurgencies (Inti-

Figure B.2
Local Religion

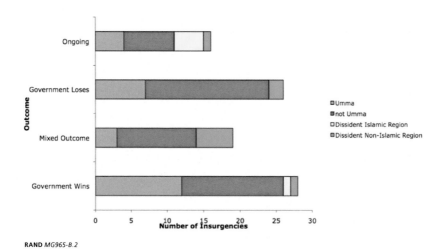

RAND *MG965-B.2*

Table B.2
Outcomes as a Function of the Local Religion

Outcome	Umma	Not Umma	Dissident Islamic Region	Dissident Non-Islamic Region
Government wins	12	14	1	1
Mixed outcome	3	11		5
Government loses	7	17		2
Ongoing	4	7	4	1

fada II, Chechnya II, South Thailand, and Kashmir on the one hand and, technically, the Niger Delta on the other).

Rate of Insurgencies

The rate at which insurgencies have started has not shown much of a trend one way or the other since World War II ended: Every two years sees roughly three insurgencies get under way (see Figure B.3).

Have insurgencies become harder or easier to win over time? Table B.3 indicates no trend in the won-lost percentage as such. What it *does* seem to show, however, is that an increasing percentage of insurgencies of late are resulting in mixed outcomes—either through explicit negotiations or through the tacit acceptance of outcomes that constitute de facto political arrangements.

Terrain

Although government was more likely to lose to an insurgency when the population was predominantly rural, terrain itself had comparatively little to do with outcomes. Governments tended to do slightly better when the terrain was less difficult (e.g., flatter), but the difference may not be significant (see Figure B.4 and Table B.4).

Figure B.3
Rate of Insurgencies

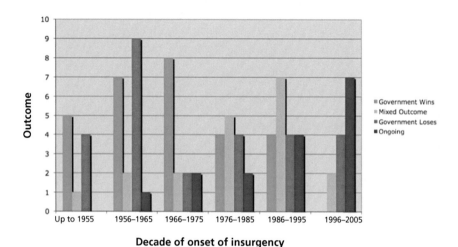

RAND *MG965-B.3*

Formulation

The outcome of any given insurgency has a lot to do with the goals sought by the insurgents. Insurgents who fought for independence or for majority rule have been almost always successful once they get going (the Mau Mau rebellion in Kenya being the notable exception) (see Figure B.5 and Table B.5). They won, in no small measure, because their campaign was consistent with the postwar zeitgeist. Conversely, insurgencies fighting for secession (or autonomy) have failed more often than they have succeeded, comporting to the principle that holds today's national borders, however arbitrarily determined, to be largely inviolable. Otherwise, the won-lost record is mixed whether the goal is establishing a Marxist or Islamic state or overthrowing the government (that is, changing the regime without necessarily changing the governing ideology).

Of note is which goals permit mixed outcomes and which do not. Such goals as independence, majority rule, Marxism, or Islamicism tend to be either-or propositions, and only four of the 34 insurgencies with such goals have resulted in a mixed outcome. Conversely, when

Table B.3
Number of Insurgencies by Decade of Onset

Outcome	Up to 1955	1956–1965	1966–1975	1976–1985	1986–1995	1996–2005
Government wins	5	7	8	4	4	0
Mixed outcome	1	2	2	5	7	2
Government loses	4	9	2	4	4	4
Ongoing		1	2	2	4	7

Figure B.4
Terrain

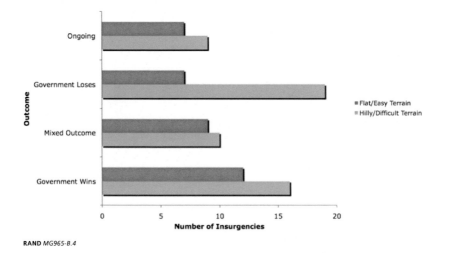

Table B.4
Number of Insurgencies by Terrain

Outcome	Hilly/Difficult Terrain	Flat/Easy Terrain
Government wins	16	12
Mixed outcome	10	9
Government loses	19	7
Ongoing	9	7

secession/autonomy or power arrangements are at issue, the difference can often be split, and mixed outcomes have characterized 15 cases, or nearly 30 percent of such insurgencies.

The difficulties that secessionist groups have of winning against an established government are made even clearer when viewed on a case-by-case basis. Of the six insurgent losses, three were in or near the Horn of Africa, and, in two of these, Somalia and Ethiopia, a region acquired its independent (Eritrea) or quasi-independent (Somaliland) status in the wake of a multi-insurgent overthrow of the central gov-

Figure B.5
Insurgency Goals

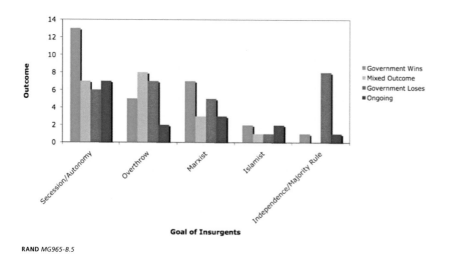

Goal of Insurgents

RAND *MG965-B.5*

Table B.5
Number of Insurgencies by Goal of Insurgents

Outcome	Secession/ Autonomy	Overthrow	Marxist	Islamist	Independence/ Majority Rule
Government wins	13	5	7	2	1
Mixed outcome	7	8	3	1	
Government loses	6	7	5	1	8
Ongoing	7	2	3	2	1

ernment.[3] The other three secessionists were clearly the beneficiaries of some major power help: Kosovo had the North Atlantic Treaty Organization (NATO) on its side; Dnistria fended off Moldova because of

---

[3] Sudan conceded an independence referendum to its southern provinces, but whether it carries through and actually allows its oil-bearing provinces to leave remains to be determined.

the support of Russia (or at least Russia's 14th Army); and Bangladesh had India to thank.[4] Among the seven mixed outcomes, three (Bosnia, Nagorno-Karabakh, and Georgia-Abkhazia) were achieved against governments that had not yet established themselves when challenged.

Finally, except for insurgencies seeking independence/majority rule, most of which started prior to 1980, almost a fifth of all other insurgencies, irrespective of goal, are still ongoing.

Start Status

It may be thought that all insurgencies start small—the proverbial ten agitators meeting in the coffee shop meeting to form a nucleus of a much larger effort down the road. Most insurgencies, do, in fact, start small and unorganized, but many of them start large and at least somewhat organized (see Figure B.6 and Table B.6). Often, the insurgents are members of prior administrations (e.g., in Iraq) or prior insurgent groups reoriented to a new conflict (e.g., UNITA in Angola). Alternatively, they may have been members of parties that never received the chance to govern (e.g., Bangladesh's Sheikh Rahman) or leaders of regional governments (e.g., Chief Ojukwu of Biafra). Some are remnants of earlier wars reactivated for a new mission (e.g., in South Vietnam). In our assessment, 52 of the 89 insurgencies started small and unorganized, while 37 started either large or at least somewhat organized. Whether or not there was a protoinsurgency, however, governments won as often as they lost. What is more striking, but not necessarily surprising, is that insurgencies that started large or organized resulted in mixed outcomes 35 percent of the time—a percentage three times as high as was true for those that started small and unorganized.

---

[4] East Timor, which was classified as a mixed outcome, was clearly helped by the international community, which was never reconciled to the 1975 absorption of Portuguese Timor into Indonesia.

**Figure B.6
Start Status**

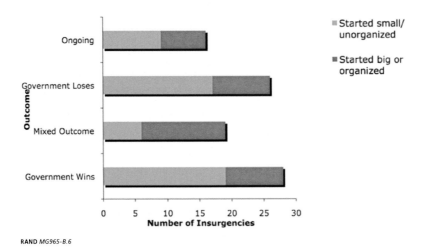

RAND *MG965-B.6*

**Table B.6
Number of Insurgencies by Start Status**

Outcome	Started Small or Unorganized	Started Big or Organized
Government wins	19	9
Mixed outcome	6	13
Government loses	17	9
Ongoing	9	7

Prewar Political Role of Insurgents

What would seem to be a simple variant on the question of protoinsurgency—did the insurgents (or at least the main insurgent) have a political role in the government prior before conflict started—turns out to be neither a variant nor particularly salient to the win-loss percentage. Insurgencies that started small tended to be led by those without a prewar role, but only by a 2:1 margin, while those with a prior role led only half the insurgencies that started large or organized (see Figure B.7 and Table B.7).

Having a prior political role also had no significant difference on who prevailed (see Table B.8).

Figure B.7
Prior Role

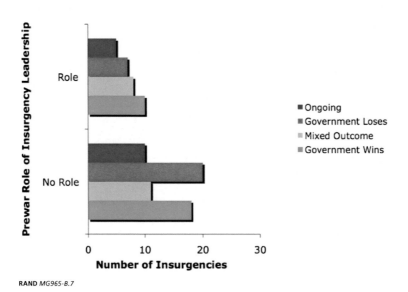

Table B.7
Correlation Between Insurgencies Whose Leaders Had a Prewar Role in the Political Process and Those That Started as Protoinsurgencies

Role	Protoinsurgency	No Protoinsurgency	Total
Prewar role	12	18	30
No prewar role	39	20	59
Total	51	38	

Table B.8
Number of Insurgencies by Prewar Role of Insurgency Leadership

Outcome	No Prewar Role	Prewar Role
Government wins	18	10
Mixed outcome	11	8
Government loses	20	7
Ongoing	10	5

Popularity

The *popularity* of the insurgency, whether one refers to the insurgents or their cause, appears to be somewhat correlated with the outcomes (see Figure B.8 and Table B.9). When the group's popularity was high (or rising), the insurgents lost only one-third of the time. When its popularity was low, it lost more than two-thirds of the time.

Figure B.8
Insurgent Popularity

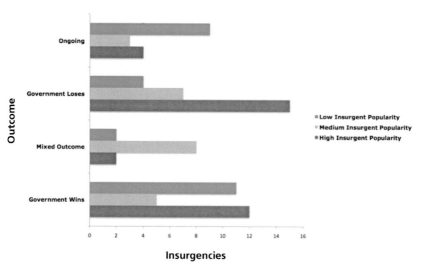

Table B.9
Number of Insurgencies by Insurgent Popularity

Outcome	High Insurgent Popularity	Medium Insurgent Popularity	Low Insurgent Popularity
Government wins	12	5	11
Mixed outcome	2	8	2
Government loses	15	7	4
Ongoing	4	3	9

This can be seen more clearly by removing the 33 insurgencies that sought secession or autonomy (see Table B.10). Because such insurgents (or at least their causes) tend to be popular within their region, the outcome of their insurgency tends to depend in part on the strength of the region compared to the nation as a whole. Overall, government has won conflicts over secession far more often than it has lost. Thus there is a large subclass of insurgencies in which the insurgents are popular but lost anyway—but this does not mean that popularity and outcome are inversely correlated.

Unity

As a general rule, the popularity of the insurgency and the popularity of the insurgents are correlated, but not always (see Figure B.9 and

Table B.10
Number of Insurgencies in Which Secession Was Not the Goal, by Insurgent Popularity

Outcome	High Insurgent Popularity	Medium Insurgent Popularity	Low Insurgent Popularity
Government wins	2	2	11
Mixed outcome	4	7	1
Government loses	10	7	4
Ongoing	1	1	6

Figure B.9
Unity

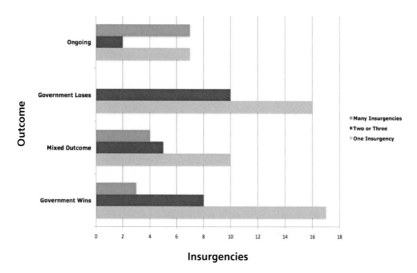

RAND *MG965-B.9*

Table B.11). Where they differ, it is often because the insurgents have, through their tactics and treatment of the population, worn out their welcome. Nevertheless, the distinction makes little difference to outcomes: The win-tie-loss distribution of those 24 cases matches those of the sample as a whole. Yet, the correlation between the popularity of the cause and the outcome is somewhat weaker than is a similar correlation between the popularity of the group and the outcome. As such, the popularity of insurgents themselves is somewhat more telling than the popularity of their cause.

Table B.11
Number of Insurgencies by Insurgent Unity

Outcome	One Insurgency	Two or Three	Many Insurgencies
Government wins	17	8	3
Mixed outcome	10	5	4
Government loses	16	10	0
Ongoing	7	2	7

Among the various forms of support for nonstate actors, it appears that getting help from fellow ethnic groups across the border is worth somewhat more to insurgents than getting help from comrades in the faith from anywhere, but the numbers may be too small in any case to permit strong conclusions.

Income

It is not a complete surprise to find that richer, more-urban countries tend to be those in which insurgents are apt to lose (see Figure B.10 and Table B.12). Among countries whose per capita incomes were less than $2,000, the government won fewer than a quarter of the time (eight of 34). This ratio rises to a third (13 of 40) among the middle-poor

Figure B.10
Income

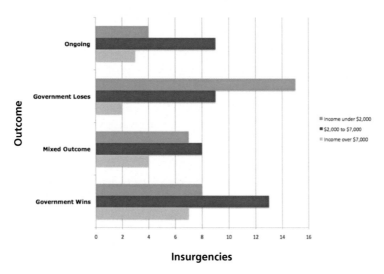

RAND *MG965-B.10*

Table B.12
Number of Insurgencies by Country Income

Outcome	Income over $7,000	$2,000 to $7,000	Income Under $2,000
Government wins	7	13	8
Mixed outcome	4	8	7
Government loses	2	9	15
Ongoing	3	9	4

countries and nearly half (seven of 16) among the more middle-income countries (which includes two for two among the rich countries).[5]

Negotiations

If insurgencies, to paraphrase Clausewitz, are politics by other means, it may be helpful to know that more than half of all insurgencies (40 out of 73) have been settled through negotiations (see Figure B.11 and Table B.13). This 40 includes all but two of the mixed-results outcomes but more than 40 percent of those outcomes that had a clear winner either way. Adding other means by which the government has recognized insurgents as other than criminals—that is, via earlier negotiations, cease-fires, or amnesty offers—encompasses all but 12 of the 73 settled insurgencies, and even three-quarters of ongoing insurgencies.

Most negotiations resulted in power-sharing arrangements (23), elections (22), or referenda (in the case of four secession movements) (see Table B.14). Some negotiations resulted in a combination of the three. Negotiations were relatively rare when the insurgents sought a religious or Marxist state but were more common when other goals were at issue.

---

[5]   O'Neill (1990) addresses income. Although our finding appeared convincing, it was also obvious and threatened to take us down a rather technical—and distracting—rabbit hole. We therefore focus on urbanization instead of addressing economics in greater depth.

Figure B.11
Negotiations

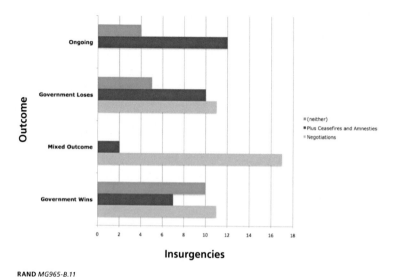

RAND *MG965-B.11*

Table B.13
Number of Insurgencies by Presence of Negotiations, Cease-Fires, and Amnesties

Outcome	Negotiations	Plus Cease-Fires and Amnesties	Neither
Government wins	11	7	10
Mixed outcome	17	2	0
Government loses	11	10	5
Ongoing	0	12	4

Presence of Foreign Soldiers

Having foreign soldiers in the ranks is a particularly interesting form of support, not least because of Iraq in particular and the various jihadist insurgencies in general. Most of the insurgencies examined have no foreign soldiers to speak of in their ranks (see Figure B.12 and Table B.15).

Table B.14
Number of Insurgencies Ended by Negotiations, by Force, and Still Ongoing, Sorted by Insurgent Goals

Goal	Negotiated	By Force	Ongoing
Independence	6	3	1
Islam	1	3	2
Marxism	4	11	3
Overthrow	12	7	3
Secession	16	10	7

Figure B.12
Foreign Forces

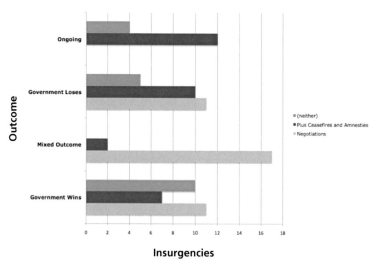

Among already-decided conflicts, they appear in only 31 of 73. However, they appear in half of all *current* conflicts. Not unexpectedly, those that have enjoyed a large number of foreign fighters have done better than average—but three of the six such victories represent the overrunning of what used to be Indochina by the armies or allies of North Vietnam.

Table B.15
Number of Insurgencies by Share of Foreign Forces

Outcome	High	Modest	Low or None
Government wins	1	6	21
Mixed outcome	1	6	12
Government loses	6	5	15
Ongoing	3	3	10

Foreign jihadists, notably those contributed by or at least associated with al Qaeda, account for most of the presence of foreign soldiers these days. Because the organization itself is of recent standing, its support is associated with only 12 conflicts, eight of which are still ongoing. Its record is 2-2 where outcomes are known (they supported the defeat of the communists in 1992 and the accession of the Taliban but were defeated with the Taliban in 2002 and can be associated with the GIA's defeat in Algeria).

Civil-Defense Patrols, Physical Barriers, and Turned Insurgents

Vigorous government-sponsored program of civil-defense patrols (e.g., Indonesia's tactic against its Darul Islam insurgency), the erection of physical barriers (e.g., Israel's wall), or the ability to turn insurgents (e.g., as the British did in Kenya) would seem to correlate with government success (see Figure B.13 and Table B.16). When all insurgencies are examined, however, the average efficacy of these tactics is less than obvious.

Incidentally, there was only one mixed outcome in the 37 cases in which the government resorted to barriers, made a policy of turning captured insurgents, or took out the insurgent leader (see Table B.17). Perhaps some actions just make compromise impossible.

Figure B.13
Civil-Defense Patrols, Barriers, and Turned Insurgents

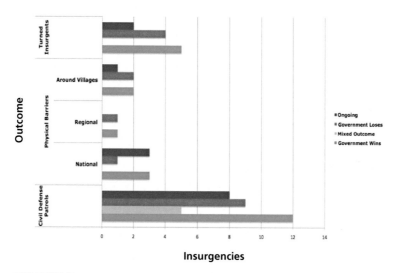

RAND MG965-B.13

Table B.16
Number of Insurgencies by the Presence of Civil Defense Patrols, Physical Barriers, and Turned Insurgents

Outcome	Civil-Defense Patrols	Physical Barriers			Turned Insurgents
		National	Regional	Around Villages	
Government wins	12	3	1	2	5
Mixed outcome	5	0	0	0	
Government loses	9	1	1	2	4
Ongoing	8	3	0	1	2

Table B.17
Number of Insurgencies by Type of Nonstate-Actor Support

Outcome	Diaspora	Coethnics	Coreligionists	Mining Interests
Government wins	1	2	3	2
Mixed outcome		2	1	
Government loses	1	3		1
Ongoing		2		1

Government Popularity

Correlation is at least equally pronounced when the subject is government rather than insurgent popularity (see Table B.18). Governments whose popularity was high or even medium won outright approximately half of the insurgencies they fought (16 out of 31 decided). But unpopular governments lost outright more than half (23 out of 42 decided).

This effect is only somewhat more pronounced when the 33 insurgencies motivated by secession or autonomy are subtracted away (see Figure B.14 and Table B.19).

Table B.18
Number of Insurgencies by Government Popularity

Outcome	High	Medium	Low
Government wins	10	6	12
Mixed outcome	2	10	7
Government loses	1	2	23
Ongoing	2	9	5

Figure B.14
Government Popularity

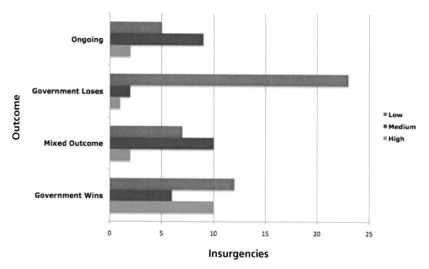

Table B.19
Number of Insurgencies Not Motivated by Secession or Autonomy, by Government Popularity

Outcome	High	Medium	Low
Government wins	8	3	4
Mixed outcome	1	7	4
Government loses	1	2	17
Ongoing	2	5	2

Strength of Government

The strength of the government (i.e., its ability to enforce its own laws and collect taxes throughout the country) is also mildly correlated in the expected direction with outcomes (see Figure B.15 and Table B.20). Oddest of all is the fact that none of the 19 insurgencies in which gov-

Figure B.15
Government Strength

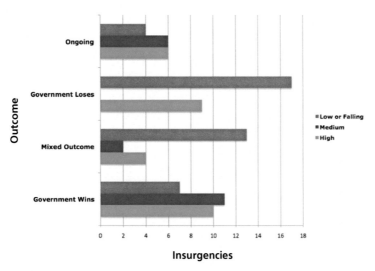

Table B.20
Number of Insurgencies by Government Strength

Outcome	High	Medium	Low or Falling
Government wins	10	11	7
Mixed outcome	4	2	13
Government loses	9		17
Ongoing	6	6	4

ernment was deemed to have medium strength was actually lost to the insurgents—but this, too, may be a statistical artifact.

Degree of Social Exclusion

Another indicator that should correlate with insurgent victory is the degree of social exclusion: The more exclusion, the greater the grievance and thus the greater the impetus, at least on the part of the excluded population, to overthrow the government. There is a correlation, but it is fairly weak and has a curious and potentially spurious angle (see Figure B.16 and Table B.21). Insurgents have won more than 40 per-

Figure B.16
Presence of Social Exclusion

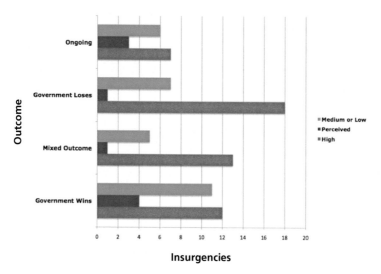

RAND *MG965-B.16*

Table B.21
Number of Insurgencies by Presence of Social Exclusion

Outcome	High	Perceived	Medium or Low
Government wins	12	4	11
Mixed outcome	13	1	5
Government loses	18	1	7
Ongoing	7	3	6

cent of all decisions (18 of 43) when social exclusion is high, but little more than a quarter (seven of 23) when social exclusion is low or just nonexistent. Oddly enough, in the nine cases in which discrimination was judged to be mostly perceived and not real, insurgents won only one.

Fear of Coup

Finally, what of the hypothesis that a government's fear of being overthrown would depress its chances of prevailing by persuading it to keep its armed forces smaller and more divided than they would be in the absence of such fears? Again, there appears to be a weak correlation between fear of coups and outcomes (see Figure B.17 and Table B.22).

Control Over Territory

What events may indicate an eventual government loss? One might nominate the insurgents' ability to hold territory at some point during the conflict (prior to victory). Yet, this was a feature of most insurgen-

Figure B.17
Fear of a Coup

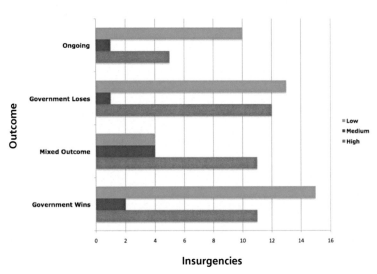

Table B.22
Number of Insurgencies by Coup Fears

Outcome	High	Medium	Low
Government wins	11	2	15
Mixed outcome	11	4	4
Government loses	12	1	13
Ongoing	5	1	10

cies, whether they won, lost, or split the difference. The holding of territory in this case is implied by their ability to prevent or inhibit government forces from entering the territory and, in most cases, the ability to extract resources from the territory in some systematic manner. Of the 74 cases in which insurgents held territory at some point, 71 involved holding regions, with the rest (Lebanon 1958, Israel, Iraq) holding only urban neighborhoods (see Figure B.18). In 18 cases, insurgents held urban areas, and, in 47 cases, there was ungoverned territory (including two cases in which insurgents held no territory of their own). Overall,

Figure B.18
Urbanization

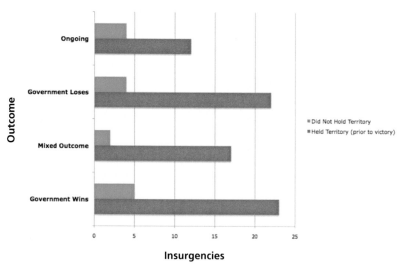

RAND *MG965-B.18*

however, there was no correlation between whether insurgents managed to hold territory prior to the end of the conflict—among those insurgencies that reached a casualty threshold—and how the conflict turned out (see Table B.23).

Elimination of Key Insurgency Leader

This study showed that counterinsurgents often attempted to achieve end state by attacking insurgent military leadership structures. In only eight of the 89 cases, however, has this tactic borne fruit (see Figure B.19 and Table B.24). The use of precision decapitation strikes came at a cost. In nearly every case where the government aggressively attacked insurgent leadership, civilian casualties or some form of civilian backlash resulted. This proved true in Iraq (2003–), Afghanistan (2001–), Turkey (PKK), Chechnya I, Uruguay (1963–1973), Northern Ireland (1969–2002), Algeria (1954–1962), South Vietnam, the Dominican Republic, and elsewhere.[6]

The ability to take out (kill or capture) a dominant *charismatic military* leader may also be helpful to governments. Insurgencies that had such a leader did somewhat better than average. However, governments that took out that leader (e.g., Abimael Guzmán of Peru's

Table B.23
Number of Insurgencies by Insurgents' Ability to Hold Territory Amnesties

Outcome	Held Territory (prior to victory)	Did Not Hold Territory
Government wins	23	5
Mixed outcome	17	2
Government loses	22	4
Ongoing	12	4

---

[6] Programs by U.S. special operations forces in Vietnam designed to kill Vietcong insurgent leaders, although successful, had a negative impact on U.S. political willpower.

Figure B.19
Elimination of Key Leader

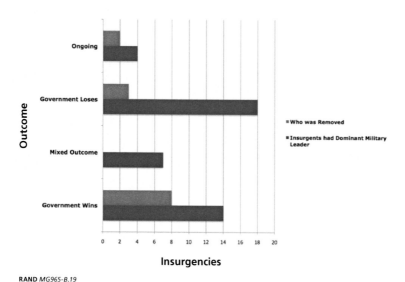

RAND *MG965-B.19*

Table B.24
Number of Insurgencies by the Presence and Removal of Dominant Military Leader

Outcome	Insurgents Had Dominant Military Leader	Who Was Removed
Government wins	14	8
Mixed outcome	7	
Government loses	18	3
Ongoing	4	2

Shining Path) prevailed more than half the time, while those that faced a dominant leader and could not or would not remove the leader won only six of 33 such insurgencies. The same is not true for taking out the dominant political leader, as many colonial or white-majority regimes did, with little success. We note that, while the capture of Guzmán in Peru helped subdue violence to manageable levels for a few years, the

Peruvian government's failure to address root causes allowed the insurgency to reignite less than a decade later.

Multivariate Regression Analysis

To explore whether broad statistics-based conclusions could be reached from the data, we ran a multivariate regression over the 73 insurgencies that had concluded. The results should be treated as interesting and suggestive rather than definitive. The primary purpose was less to "explain" why insurgencies were won or lost (or tied) than to determine the common features shared by insurgencies that were won (by the government), were lost, or resulted in a mixed outcome. We further note the small sample size, the large number of potential explanatory variables, and the necessary use of judgment calls in coding and evaluating them. In addition, the use of cardinal measures to represent an ordinal dependent variable—outcome—may be considered nonstandard. We judged a government victory equivalent to 3; a mixed outcome, 2; and an insurgent victory, 0. But insurgency is not ice hockey; a mixed outcome reflects not only a rough equivalence between forces (in that neither side can eliminate the other) but willingness on both sides to settle rather than press on. The most likely alternative to a mixed outcome is often continued warfare.

We found the best results from 11 independent variables[1] representing eight parameters (insurgent motives were coded as being

---

[1] Many of the variables, such as government popularity and strength or insurgent structure, are the same as reported earlier. Others are somewhat different. For insurgent terror, broad terror was coded as 2, no terror as 0, and everything else as 1. As noted in the text, insurgent goals and government type were coded in terms of dummy variables. The force-ratio variable was constructed by comparing government and insurgent forces: 3 if the government had at least a 10:1 force ratio, 2 if the force ratio was between 3:1 and 10:1, 1 if the force ratio was between 1:1 and 3:1, and 0 if the insurgents had more forces. Urbanization

independence/majority rule, secession, or other; government type was coded as democracy, anocracy, autocracy, or other). Omitted were variables with little discernable explanatory power, as well as those associated with outsider intervention (since such forces were involved in only 21 insurgencies).

The equations yielded a respectable R^2 of 0.61 with an adjusted R^2 of 0.53.[2] The total equation is represented by Table C.1.

For a variety of reasons, notably the tendency for many of the variables to correlate with one another, only two variables were statisti-

Table C.1
Explanatory Factors That May Influence the Outcome of Insurgencies

Outcome	Coefficient	Standard Error
Intercept	3.1202	0.8361
Independence	−1.0697	0.6617
Secession	0.3261	0.2597
Structure	−0.1075	0.1295
State support for insurgents	−0.5896	0.1479
Insurgent terror	0.3005	0.1339
Sanctuary (index)	−0.2564	0.1292
Government popularity	0.2850	0.1758
Government strength	0.2189	0.1578
Democracy	0.0963	0.6593
Autocracy	−0.1575	0.6417
Anocracy	−0.8628	0.5931
Urban/income (index)	0.1450	0.1198

and income levels resulted from combining the two highly correlated variables; the value ranged from 0 (least urbanized/poor) to 5 (most urbanized/middle class). The state-support variable is 2 for insurgencies that enjoyed such support, 1 for those that did not, and 0 for those whose support was withdrawn before the conflict ended. Similarly, the sanctuary index is 2 for insurgents that enjoyed voluntary sanctuary across the border, 1 for those that had sanctuary that was not voluntarily provided by the neighboring government, and 0 for those with no sanctuary.

[2]   The adjusted R^2 discounts for the number of variables introduced into the equation. One can always get a better R^2 by simply adding irrelevant variables.

cally significant at the 95-percent level: (1) state support for insurgents and (2) the insurgent's use of terrorism.

Another method we used to evaluate the data was to create an amalgamated factor score for each insurgency (see Table C.2). A score represents the tendency of a given insurgency to be decided in the government's favor, in the insurgent's favor, or with a mixed outcome.[3] Not surprisingly, each insurgency that is won by the government tends to share characteristics associated with other insurgencies that are won by governments. Using such a measure allows one to forecast the ultimate outcome of current insurgencies. One examines the relevant factors in undecided insurgencies to generate an amalgamated factor score. If the factors associated with government victories in the past apply to the future, then this score would suggest how confidently one could predict an ultimate government victory, government defeat, or mixed outcome.

When the amalgamated factor score is at least 2.20, the government wins more than 75 percent of the time and never loses outright. When the score is below 1.18, insurgents win more than 80 percent of the time and never lose outright. Scoring the 16 ongoing insurgencies by these criteria suggests that the government is likely to win 14 of them and have to settle for a mixed outcome in only two cases. Time will tell, however, if the variables that described past outcomes describe outcomes yet to ensue.

The distribution of scores sorted by outcome (together with the vertical zone bars) is presented in Figure C.1.

---

[3]   The equation is (3.1202 − 1.0697)(Is_Independence + 0.3261)(Is_Secession − 0.1075) (Structure − 0.5896)(State_Support_for_Insurgents + 0.3005)(Insurgent_Terror − 0.2564)(Sanctuary_Index + 0.285)(Government_Popularity + 0.2189)(Government_Strength + 0.0963)(Is_Democracy − 0.1575)(Is_Autocracy − 0.8628)(Is_Anocracy + 0.145) (Urban/Income_Index). One generates an amalgamated score by plugging in the value of the various factors for each insurgency.

Table C.2
Insurgencies by Regression Score

Score	Government Wins	Mixed Outcome	Insurgents Win	Ongoing
Over 2.20	22	6		14
1.18 to 2.20	6	10	5	2
Under 1.18		3	21	

Figure C.1
Scores

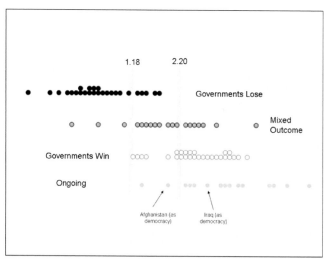

Insurgencies Not Examined for This Publication

The following is a list of 51 insurgencies that appeared on the Fearon-Laitin list but were not analyzed in this project. We tended to omit conflicts that would shed little light on our key question of how insurgencies end. Many of these are closer in nature to civil disturbances, coups, coup attempts, or spontaneous insurrections. Of the 51, almost half (25) did not see their second birthday, and all but nine of them ended within six years. Eighteen of them had begun before 1956.

Table D.1
Insurgencies Not Examined for This Project

Insurgency	Years
Costa Rica	1948–1948
Haiti	1991–1995
Bolivia	1952–1952
Paraguay	1947–1947
Argentina	1955–1955
Indonesia	1945–1946
Rwanda	1956–1961
Madagascar	1947–1948
Tunisia	1952–1954
Morocco	1953–1956
Cameroon	1955–1960

Table D.1—Continued

Insurgency	Years
Yugoslavia	1991–1991
Cyprus	1974–1974
USSR Estonia	1946–1948
USSR Lithuania	1946–1950
USSR Ukraine	1946–1950
USSR Latvia	1946–1947
Guinea	1998–1999
Mali	1989–1994
Chad	1965–
Chad	1994–1998
Congo (Brazzaville)	1998–1999
Congo FLNC	1977–1978
Uganda NRA	1981–1987
Pakistan (Sindhis versus Mahajirs)	1993–1999
Burundi	1972–1972
Burundi	1988–1988
Rwanda	1962–1965
Djibouti	1991–1994
Ethiopia ALF	1997–
Angola-Cabinda	1992–
Zimbabwe-Ndebele	1983–1987
Algeria: Kabylie	1962–1963
Sudan Anya Nya	1963–1972
Iran Kurdistan	1979–1993
Turkey militias	1977–1980
Iraq Shamar	1959–1959

Table D.1—Continued

Insurgency	Years
Yemen	1948–1948
Yemen 1994	1994–1994
Yemen 1986	1986–1987
Iran Khomenie	1978–1979
South Korea	1949–1950
India Sikh	1982–1993
Bangladesh	1976–1997
Sri Lanka JVP I	1971–1971
Sri Lanka JVP II	1987–1989
Moloccas	1950–1950
Indonesia W. Papua	1965–
Xinjiang	1991–
Brazil	1965–1972

NOTE: USSR = Union of Soviet Socialist Republics.
FLNC = Front for the National Liberation of the
Congo. NRA = National Resistance Army. ALF = Afar
Liberation Front. JVP = Janatha Vimukthi Peramuna,
or People's Liberation Front.

Categories Used for the Spring 2006 Survey

Various RAND analysts and research assistants—most of whom had enough knowledge of the region or insurgency to reach conclusions on their character based on earlier research—were asked to evaluate, for each insurgency, a set of variables: those pertaining to the insurgents, those pertaining to the government, and those pertaining to countries that intervened on behalf of the government. In addition, each insurgency was accompanied by a narrative description of the insurgency's course and outcome to validate or at least put into context the variable descriptions. Further data were provided on when the insurgency started (based on when the insurgency organization was put together) and when it ended (based also on political considerations). Finally, the income, urbanization, terrain, and religion of each country were coded.

Variables Associated with the Insurgents

Many characteristics of the insurgents themselves were tested as having some useful correlation with the conflict's outcome. Where there were multiple insurgent groups, an attempt was made to assess these variables over all of them. When impossible or meaningless, the largest group was used as a reference point. The following attributes were coded:

- size: the number of active insurgent fighters at its peak[1]
- competence at insurgency: how good the insurgents were at carrying out insurgency, a qualitative measurement[2]
- group popularity: whether the group was popular (in the sense of whether they could have won election) *in the region for which it is fighting*[3]
- cause popularity: an indicator of a cause that was more popular than the insurgents fighting on its behalf (e.g., because the insurgents were excessively bloody-minded)
- goal: what the insurgents were trying to achieve (where insurgents had multiple goals, such as independence and Marxism, one primary goal was selected)
- structure: whether the C2 for the primary insurgent group was tight and hierarchical (clearly enforced chains of command) or looser and, in a sense, more networked (cells enjoy quasi-autonomy in theory)[4]
- unity: a variable to indicate how many different insurgent groups were fighting: one, a few, or many
- international support: states or nonstate actors (e.g., Islamists, a diaspora) that provided material and even manpower support to the insurgency, if such support was of material significance

---

[1]   Wide variations in estimates are common. When Savimbi's UNITA insurgency ended in 2002, the Angolan government offered a compensated amnesty for former insurgents. The number who showed up to claim such status proved to be far higher than the number earlier estimated to be the population of insurgents.

[2]   Where the insurgency, as such, was fought by primarily conventional means, this variable would be indicated as "N/A."

[3]   This last clause relates to insurgencies whose goal was regional secession. For instance, the PKK's Kurdish insurgents could probably not care less about winning a nationwide election (e.g., throughout Turkey); what mattered was the ability to garner support from their region (e.g., to win a regional election against a propeace party).

[4]   Marxist insurgents were typically coded as hierarchical largely because such an organizational scheme fits the overall ideology (e.g., "dictatorship of the proletariat"), even when cadres (e.g., in the Vietcong) had effective autonomy in the sense of being able to operate, out of necessity, without constant communications with headquarters.

- percent foreign: the share of the insurgent's fighters that came from somewhere else[5]
- sanctuary: the adjoining country or countries, if any, to which the insurgents could retreat[6]
- al Qaeda: whether al Qaeda supported the insurgency
- terror: whether the insurgents used terror, and, if so, whether the terror was broad (random civilians were targeted), atrocities (random civilians died at the hands of both the insurgents and the government), discrete (e.g., victims were almost always members of the government), or generally not used.

Variables Associated with the Government

Many variables associated with the government resembled those associated with the insurgents:

- size: how many fighters the government had[7]
- competence at COIN: how good the government was operationally
- popularity: whether the government could win a free election against the insurgents in the country or in the insurgent region

---

[5] Because local refugee camps, for this purpose, are not considered somewhere else, PLO participation in Jordan (1970) or Lebanon (1975–1990) was not considered foreign, as such. However, soldiers of another country's military (e.g., South Africa's in Angola) were considered foreign.

[6] The sheltering country did not have to support the insurgency itself. In some cases, it sufficed if parts of the government did (e.g., Pakistan's Inter-Services Intelligence [ISI] is thought be helping the Taliban); general sympathy also sufficed (e.g., the role Turkish Kurdistan played for Iraq's Kurds). Sometimes (especially in Africa), the providing country may have lacked the strength to keep the insurgents out; whether the sanctuary was provided voluntarily was asked in the autumn 2006 survey.

[7] In some cases, measuring the size of the army sufficed. When the insurgency was well localized, only those soldiers deployed to the area were counted. Where government forces included soldiers, police, home guard, or even freelance militias (e.g., Sudan's janjaweed) or home guards, some approximate accommodation for these forces were made.

- strength: the extent to which the government could exert its will—enforce laws, collect taxes—throughout the country, or, if relevant, within the insurgent region[8]
- type: e.g., colonial government, autocracy (includes communism, Islamicism, and monarchy) democracy, and anocracy (nominal democracy without free and fair elections coupled with rule of law)
- exclusion: whether specific ethnic or religious groups or classes exercised disproportionate political power[9]
- coup: whether the government feared being overthrown by its military in a coup.[10]

Variables Associated with the Outside Intervention Forces

Only 29 of 89 insurgencies featured significant outside intervention; of these, nine were solely indirect, leaving "blue" forces to be analyzed only for the 20 cases in which such forces were on hand. The following variables are defined for the major blue force only (e.g., in Vietnam, U.S. forces but not South Korean forces):

- who: which country, countries, or multinational organization
- direct: if combat units are sent, the assistance is direct; otherwise, it is indirect
- size: the size of the primary military contingent at its height

---

[8] The question of strength is often a circular one: A strong insurgency means a weak government even if the government is inherently strong in the sense that it was strong in the insurgent region prior to the insurgency and strong elsewhere in the country. Here, the intent is to measure inherent strength, not temporary reductions in it.

[9] If the insurgents' supporters believe that they are excluded despite strong evidence that they are not, this variable is coded as "perceived."

[10] The question was raised to test the hypothesis that a timorous government would be therefore likely to keep the army weak (e.g., small size, operational restrictions, fractionated C2) and thus incapable of pursuing COIN with requisite energy and resources.

- competence at COIN: how good the intervener's forces were at conducting COIN
- popularity: whether the interveners were basically welcomed or not
- ab initio: whether the interveners were in country from the beginning, arrived shortly thereafter, or well after the fighting started.

Variables Describing the Country

Four variables were used to describe the country in which the insurgency was taking place (these answers were looked up by a research assistant or the author and were not posed to the analysts):

- income: measured in purchasing-power-parity dollars according to the most recent World Bank estimates (for analytic purposes, the three ranges were under $2,000/year, $2,000 to $7,000/year, and over $7,000/year)
- percent urban: the percent of the population (of the overall country) living in towns and cities (for analytic purposes, the three ranges were under 40 percent, 40 to 70 percent, and over 70 percent)
- terrain: how mountainous the terrain is
- religion: whether the country or region was majority Islamic.[11]

---

[11] How the answers were coded depends on whether the insurgency in question aims at secession. If not, then the variable is either yes or no. If the insurgency is an attempt at secession, then four answers are possible: yes (the country and the region are majority-Islamic); no (neither the country nor region is majority-Islamic); yes-region (the region is majority-Islamic but the country is not); or no-region (the country is majority-Islamic but the region is not).

Unavoidable Ambiguities

Our analysis presumed that several distinct actors could be identified in a typical insurgency: the insurgents, the local government, and, in some cases, forces from outside that come to the aid of the government (in color parlance, red, green, and blue, respectively).[1] Although most insurgencies did not have blue actors, the roughly one-quarter that did are of particular note, since they tend to provide an informative precedent for U.S. participation in current or future conflicts.

In practice, insurgencies constitute a variegated lot, and making clean definitions of red, green, and blue is anything but straightforward.

Some insurgencies, for instance, are led by a single entity (e.g., the Vietcong). Others are carried out by multiple groups of insurgents, with often-differing goals and coordinated more or less well with one another. It is not unknown for the various insurgents to turn on and fight each other after the government has been overthrown. The postindependence struggle in Angola between MPLA (which formed the new government) and Jonas Savimbi's UNITA forces dragged out more than twice as long as the original preindependence insurgency.

In most cases, the government was unique and easy to define (for colonies, it was defined as the imperial ruler). But there were still some ambiguities. Some insurgencies (e.g., Lebanon, 1975–1990; Colombia, 1948–1962) degenerated into intercommunal violence in which

---

[1] A few insurgencies, such as the Northern Alliance versus the Taliban and UNITA versus the Angolan government, were assisted by the regular armed forces of another state. Rather than generate a separate color for those forces, such instances were noted in the variable that dealt with international support for the insurgents.

the government played no real role. Since 1990, warlord-dominated Somalia has had no government to speak of; the June 2006 ascendancy of Islamists in Mogadishu, Somalia, left the country with two quasi-official governments for the next half-year. Over most of the past 25 years, Afghanistan's government could only be defined as whoever occupied Kabul.[2] In four cases (Yemen, Bosnia, Sierra Leone, and Liberia), identifying which group should be counted as the government was a judgment call: It was mostly a question of who controlled the capital city at any one point in time. Finally, in some insurgencies, the character of the government itself changed radically (e.g., from autocracy to democracy) over the course of conflict. Nevertheless, in 81 of the 89 cases, the established government was unambiguously a party to the conflict.

Defining blue was not always easy either. Where one state lent troops to another, at least a primary identification was straightforward; less so, when the assistance was indirect. In at least three cases, blue consisted of peacemakers who ended up fighting the insurgents (the UN in Katanga, India in Sri Lanka, and Economic Community Military Observation Group [ECOMOG] in Liberia). We elected not to color those peacemakers who did not fight the insurgents as blue.

Determining the outcome of the insurgency is critical to any analysis of how government-won insurgencies differed from those insurgents won or from those that yielded mixed outcomes. Where the insurgency disappeared without altering the government, or where the insurgents took over the government on their own, the outcome was clear cut. Conversely, many insurgencies ended in negotiations or tacit cease-fires, requiring an assessment of whether enough of the insurgent's aims had been achieved to qualify as a mixed outcome or whether it really was a government victory. In some cases, outcomes had little to do with the insurgency per se but were instead imposed by the outside (e.g., East Timor's independence); coding such outcomes

---

[2] We coded four separate Afghan insurgencies in the past quarter-century: against the Soviet-backed government; the years of chaos that culminated in the Taliban's taking power; the insurgency that, with U.S. help, toppled the Taliban; and today's fighting in which the Taliban are, again, the insurgents.

as victories for the insurgents would not say anything meaningful about the actual qualities of the insurgency or COIN. Several insurgents achieved de facto autonomy for their region; these were coded as victories. Finally, there had to be some determination of when an insurgency had produced an outcome. In 16 of the 89 cases, it had not. Some (e.g., Iraq) are clearly ongoing. Others (e.g., Chechnya) could be evaluated the other way.

Government Victory

As mentioned above, most violent internal conflicts end in state victory. States typically ignore insurgencies in their nascent stages, dismissing the antigovernment opposition as mere "bandits" or "terrorists." Acknowledging the unrest as an insurgency confers a politically important element of legitimacy on the opposition. What is more, such recognition is a de facto admission that something was terribly wrong with the polity—perhaps fatally so.

Once the threat is finally recognized and acknowledged, anti-insurgent actions can take a variety of forms. In "classic" Western COIN, the incumbent government, using political, military, economic, intelligence, and other instruments of national power, gradually wins popular allegiance through reforms and other blandishments, establishes security, and neutralizes the insurgent political and military structures.[3] Malaya (1948–1960) remains the canonical COIN success.

Classical COIN is, of course, not the only means by which incumbents have succeeded in defeating internal armed opposition. Although distasteful to liberal democracies (if sometimes employed by them), extreme repression can be remarkably effective in destroying insurgent groups, as demonstrated in Algeria during the 1992–1998 period. In its ferocious campaign against the GIA, the Algerian army adopted a policy of "terrorize the terrorist," responding to each fresh terrorist outrage "blow for blow" (Benramdane, 1999). This took the form of extra-

---

[3]   For an overview, see Blaufarb (1977). For a still-useful critique of U.S. COIN strategy, see Schwarz (1991).

judicial killings of suspected GIA militants and their supporters, the terrorization of populations suspected of radical Islamist sympathies, and a program to "make fear change sides" by establishing self-defense forces across the countryside and using them in offensive and defensive roles against the insurgents.

Other Than Government Victory: Loss, Mixed, Ongoing

Instead, insurgencies most often end because one side loses essential external support (e.g., the withdrawal of Iranian support for the Kurdish insurgency in Iraq in 1974), the intervention of outside military forces (e.g., the Syrian invasion of Lebanon in 1976) or, more frequently, because one of the warring parties wins by military means.

The diversity of insurgencies makes generalizations difficult (Laqueur, 1976, p. 386). Insurgencies differ in terms of motivation (e.g., revolutionary, ethnic-nationalist, salafist-jihadist), leadership (charismatic, collective, cellular), and strategy ("people's war," attritional, insurrectional). Violent internal conflict can be protracted, as in Burma, where separatist guerrillas have fought the state (and each other) for nearly six decades, or relatively brief, as in Cuba during the 1950s, when insurgents gained power after only a few years of sustained fighting. Other important features, such as the level of popular support, the competence of COIN forces, and the role of external actors, can vary widely from conflict to conflict.

But with respect to end games and outcomes, much less variety is evident. Although much of the scholarly and policy-oriented research focuses on negotiated outcomes, insurgencies seldom draw to a close as a result of what takes place at the bargaining table. In the vast majority of internal conflicts, the incumbent prevails, and typically does so in a decisive way. Insurgent victories are rarer, and, when they do occur, they frequently contribute to shifts in the tectonic plates of international politics. In some instances, however, deciding who is the victor and who is the vanquished presents analytical quandaries, as in the case of colonial authorities in Kenya who crushed the Mau Mau rebellion but shortly thereafter relinquished control over Kenya.

Insurgents lose slowly; as their effectiveness declines, it does so at a decreasing rate. Governments, on the other hand, lose more quickly; the end comes with a dramatic thud as the state collapses on its own disintegrating foundation. How can observers and practitioners detect whether the conflict's end game is under way? As the end approaches, people "vote with their feet" in hurried attempts to avoid being on the wrong side of the struggle as it draws to a close, thus creating a "negative bandwagon" effect. Government defeat is telegraphed by increasing rates of defections and desertions, particularly among senior officers and civilian leaders; capital flight; and the drying up of popular sources of actionable intelligence. Similarly, the end game for insurgents can be signaled by a growing tempo of departures from the resistance movement. Additionally, the elimination of internal or external safe havens, and a substantial reduction in external support from diasporas or other governments, are likely to prove lethal to the insurgency and are thus key indicators that the end has begun—or is likely to begin soon.

Five other notable inferences about how insurgencies end can be drawn from the combination of the quantitative and qualitative studies. They address the most-significant factors appearing in both kinds of research: the role of external support; fragmentation of power; extreme violence; defection and infiltration; and local security forces. A brief discussion of the first two inferences follows; the latter three were discussed in the main text.

External Support

First, external support for either the insurgents or the government or both has major effects on the patterns of conflict termination (or lack thereof). Insurgencies that receive external support or sanctuary in a foreign country are much more successful. Conversely, even governments that appear to be weak and ripe for collapse can stave off insurgent victory with significant external support. This conclusion is among the most statistically robust from the quantitative study.

The importance of external support is also clearly demonstrated in the Afghanistan and Lebanon cases and, to a lesser extent, in the Sri Lanka and Northern Ireland cases. In Afghanistan, external support was important to both government and insurgents. Even after the

Soviet troop withdrawal, continued military and financial aid allowed the communist regime to cling to power for more than two years. The Afghan insurgents benefited extensively from both material support from abroad and sanctuary in Pakistan. In Lebanon, Syria's influence on the insurgency was pervasive. In Sri Lanka, the Tamil diaspora provides important funding to the insurgency, as did the Irish diaspora for the Irish Republican Army (IRA) in Northern Ireland.

Fragmentation of Power

Second, conflict termination of any type (though especially negotiation) is generally made much more difficult by the presence of more than two centers of power. Put another way, conflict termination is difficult enough when it involves only a unitary government and a unitary insurgent group. When the government is not unitary, the insurgency is not unitary, or a third force exists (such as paramilitaries independent of the government), conflict termination becomes vastly more difficult. This is not particularly surprising, but it is a common enough pattern that evaluation of the fragmentation of power should be explicitly included in evaluations of insurgency.

This pattern is prominent in the Kenya, Afghanistan, Northern Ireland, and Lebanon cases. In Kenya, white settlers and their loyalist Kenyan supporters had goals and methods that were often at odds with the British government. In Afghanistan, the fragmented insurgency made a conclusive ending to the insurgency almost impossible, resulting in a subsequent civil war. In Northern Ireland, Protestant paramilitaries had to be tamed before the Catholic insurgency could be ended. In Lebanon, the mix of Druze, Sunni, Shi'a and Christian made an end to the conflict both difficult and unstable, as subsequent violence has shown.

Questions Used for the Autumn 2006 Survey

The following questions were given to each analyst (the answers to questions 21 through 23 were generated by the author and then amended as necessary by each of the analysts).

1. Is the conflict best described as (a) an insurgency against an established government, (b) a conflict among several groups where there is no effective government, or (c) a conflict among several groups one or more of which variously occupy the position of government? **Answer = a, b, or c**

2. Was the conflict settled through negotiated settlements? **Answer = yes or no**

   a. If yes, did such settlements call for elections (i.e., thereby legitimizing the insurgents as a political party)? **Answer = blank, yes, or no (for some secessionist insurgencies, a third answer, "referendum" was possible)**

   b. If yes, did they redistribute power within the national government (e.g., give cabinet positions to insurgents)? **Answer = blank, yes, or no**

3. Were there negotiated settlements prior to the one that ended the conflict? **Answer = yes or no**

   a. If yes, why/how did they fail to end the conflict (e.g., what broke down)? **Answer = blank, or text (explanation—which was then grouped into three categories: government's fault, insurgents' fault, both at fault)**

   b. If yes, were there splinter groups that carried on the fight? **Answer = blank, yes, or no**

4. Were cease-fires agreed on and broken prior to the end of conflict? **Answer = yes or no**
5. Did outside (that is, largely disinterested or humanitarian) intervention play a key role in settling the conflict? **Answer = yes or no**
   a. If yes, did it exert primarily military pressure, as opposed to other forms of pressure? **Answer = blank, yes (primarily military), or no (other forms)**
   b. Did it offer credible guarantees? **Answer = blank, yes, or no**
   c. Did it involve peacekeeper forces? **Answer = blank, yes, or no**
6. Did the resolution of the insurgency follow from a(n unexpected) change in the government (via, e.g., an election or coup)? **Answer = yes or no**
7. Did the government offer the insurgents amnesty as part of its resolution strategy? **Answer = yes or no**
8. Did the support of nonstate actors make a material difference in the nature or timing of the outcome? **Answer = yes (there was support, green did not win, and such support made a difference to the outcome) or no**
   a. If so, was such support from the diaspora in noncontiguous countries? **Answer = blank, yes, or no**
   b. If so, was such support from neighboring coethnics in neighboring countries? **Answer = blank, yes, or no**
   c. If so, was such support from religious groups? **Answer = blank, yes, or no**
9. Was there a single powerful (and omnipresent) *military* leader in charge of the insurgents? **Answer = yes or no**
   a. Was this leader removed? **Answer = blank, yes, or no (where the dominant *political* leader was arrested, this was separately noted)**
   b. Did it doom the insurgency? **Answer = blank, yes, or no**
10. Was there state sponsorship of the insurgency? **Answer = yes or no**
    a. Was state sponsorship ended? **Answer = blank, yes, or no**

b. If state sponsorship ended, did the state sponsor switch sides (vice stop supporting one side)? **Answer = blank, yes, or no**

c. If state sponsorship ended, did this doom the insurgency? **Answer = blank, yes, or no**

11. Did the conflict involve sanctuary in another country? **Answer = yes or no**

a. If so, was it voluntary or involuntary on the part of the sanctuary provider? **Answer = blank, yes (voluntary), or no (involuntary)**

12. Were walls, berms, or fences erected to control insurgent movements into and out of the country? **Answer = blank, yes, or no**

a. If so, was it (a) all of the country, (b) a large part of the country, or (c) selected districts in the country (similar to those the British used in the Boer War)? **Answer = a, b, or c**

b. If so, did they make a significant difference to the ultimate outcome? **Answer = blank, yes, or no**

13. Did the insurgent's military forces suffer conventional military defeat on the battlefield? **Answer = yes or no**

a. If yes and the insurgency lost, was the battlefield loss a relevant factor in its defeat? **Answer = blank, yes, or no**

14. If the government lost:

a. Was it the result of military defeat? **Answer = blank, yes, or no**

b. Was there a more-or-less obvious withdrawal of popular legitimacy? **Answer = blank, yes, or no**

c. Were there defections from the government? **Answer = blank, yes, or no**

d. Was there a rapid build-up of insurgent forces prior to the end? **Answer = blank, yes, or no**

e. Did other insurgent groups arise to help topple the government at the end? **Answer = blank, yes, or no**

15. Was the insurgency treated as the government's primary problem (as indicated by the percentage of its armed forces that was committed to the conflict) or did it appear that the government considered it a regional affair of less importance? **Answer = yes or no (sometimes a mixed value was indicated)**

16. Did the insurgents ever control territory (at least significantly prior to their becoming the established government)? **Answer = yes or no**
    a. If so, was it regional? **Answer = blank, yes, or no**
    b. If so, was it just in selected urban neighborhoods? **Answer = blank, yes, or no (a yes answer was permissible only if the answer to 16a was no)**
    c. In any case, were there areas of the country that were effectively ungoverned? **Answer = yes or no**
17. Did the government establish civil(ian) defense patrols? **Answer = yes or no**
18. Did the government use "turned" insurgents in pseudo units? **Answer = yes or no**
19. Did the insurgents ever have a legitimate or quasi-legitimate role in the political process prior to or in the early stages of the insurgency? **Answer = yes or no**
20. Can the last year of the insurgency be determined unambiguously (as opposed to a gradual petering out or stalemating of the insurgency over time)? **Answer = yes or no**
21. What did the last year of the insurgency look like? **Answer = blank** (ongoing)**, ambiguous** (no distinct end date)**, under a year** (the insurgency was that short)**, conventional** (warfare consisted solely of conventional warfare)**, new government leader, removed insurgent leader, end of Cold War, intervention, unintervention** (withdrawal of active state support from the insurgent), and **a culminating last year.**
22. Was there a protoinsurgency phase (e.g., did the insurgents start off small and unorganized)? **Answer = yes or no**
23. Was the national capital under threat (at least to the extent that life there was made much more difficult)? **Answer = yes or no**

Glossary

anocracy. For the purposes of this study, *anocracy* is defined as a government that is a democracy in name only. In an anocracy, the government claims to support various democratic principles while ensuring that traditional autocratic power structures remain dominant, if not obviously so. The populace has little faith in the government or in the rule of law (Gurr, 1974; Vreeland, 2008).

arbakai. A historical volunteer militia model found in parts of rural Afghanistan. *Arbakai* may be translated from Pashtu to mean "guardian."

causation. This is a provable relationship between cause and effect. An action or event is causative if it can be clearly traced to an outcome. An event can have a single identifiable cause or multiple causes. In the case of insurgency, proving causation is very difficult, and proving one action or event to be solely causative is nearly impossible.

correlation. This is a concurrence of actions and events that may or may not be causative but that are, in some way, related. Correlation is easier to show than causation. Correlation is typically shown through statistical analysis.

counterinsurgent. The term *counterinsurgent(s)* is herein used to mean government, government forces, and external intervention forces fighting on behalf of the government and against the insurgents and their sponsors.

firqa, firqat. *Firqa* is a transliterated Arabic word for team or military unit. In many cases, the plural *firqat* is often used in place of the singular *firqa*. The transliterated Arabic plural is then Anglicized for plural usage as *firqats*.

foco, foquismo. *Foco* is Spanish for "focal point" or "nucleus." Communist insurgent strategist Ernesto "Che" Guevara and a contemporary, Régis Debray, articulated the belief that a small but energetic and morally guided insurgent group could spark a revolutionary passion among the rural poor. The insurgency would then quickly blossom and, in the best-case scenario, lead directly to government collapse. This concept differs from Mao Tse-tung's insurgency theories in that it offers the possibility of spontaneous uprising rather than a purposefully long war. The term is typically associated with communist or communist-inspired insurgency movements in Latin America in the 1960s and early 1970s, none of which bore fruit. Guevara himself was killed in a failed attempt to ignite a *foco* insurgency in Bolivia. *Foquismo* is a descriptor.

hibernation. The insurgency has either suffered a serious military blow, lost popular support, or both, and has effectively dissolved as a fighting force. The insurgent organization either reverts to the protoinsurgency phase or breaks apart, forming new protoinsurgency splinter groups that may take years or decades to foment. In most cases, hibernation reflects the simple fact that the government has failed to address the underlying causes of the insurgency.

insurgency. The public debate on Iraq clearly revealed that one person's insurgency is another's civil war. This monograph does not intend to settle scholarly disputes over the differences between insurgency, revolution, terrorism, rebellion, civil war, and civil unrest. For this study, *insurgency* is taken at its broadest definition. In our review of counterinsurgency literature, we found that many experts took a similar approach. Briefly, insurgency is the violent struggle by a nongovernmental armed group against its government or an interceding force, with the intent of overthrowing the current regime, expelling an interloper, gaining greater rights, or obtaining independence. The terms *guerrilla* and *revolutionary* are folded into this definition, while, in other studies, *guerrillas* are separated from *supporters* and *logisticians*. In the cases in which foreign fighters have joined insurgent groups, they are described collectively with indigenous forces unless otherwise noted. We generally address terrorism as a tactic and draw a subjective distinction between insurgent and terrorist organizations. For example, while we studied the Provincial Irish Republican Army as an insurgency, we viewed the splinter Real Irish Republican Army as a terrorist organization and did not include it in our data set.

jazira. Modern standard Arabic for "island" or "peninsula." For our purposes, the term *jazira* refers to isolated areas in the Iraqi desert used as internal sanctuary by various insurgent groups.

protoinsurgency. This is an insurgency in its nascent or infant state. Protoinsurgents typically have no military structure, but they may be capable of some limited action. Lacking cohesion, support, and organization, they present little real threat to the government, at least until they develop into full-fledged insurgencies (unless one subscribes to the concept of foquismo). Because this study focused on insurgency endings, protoinsurgencies were not included in the selected database. Daniel Byman (2007) defines and explores the life cycle of the protoinsurgency in the RAND occasional paper *Understanding Proto-Insurgencies*.

sanctuary, safe haven. A secure or relatively secure area from which insurgencies can command, control, operate, rest, refit, and train. There are three basic forms of sanctuary: voluntary external, involuntary external, and internal. External sanctuary is typically provided by a neighboring state or territory. External sanctuary is provided knowingly and with permission and support, is provided unwittingly, or is allowed only involuntarily. *Unwitting* is equated with *involuntary*. Internal sanctuary is a relatively safe or inaccessible area within the borders of the country or territory in which the insurgency operates.

Bibliography

"200K Said Killed in Algeria Insurgency," ABC News International, March 18, 2006.

Abraham, Thomas, "The Emergence of the LTTE and the Indo-Sri Lankan Agreement of 1987," in Kumar Rupesinghe, ed., *Negotiating Peace in Sri Lanka: Efforts, Failures and Lessons*, London: International Alert, 1998.

AbuKalil, As'ad, "Government and Politics," in Thomas Collelo and Harvey Henry Smith, eds., *Lebanon: A Country Study*, 3rd ed., Washington, D.C., Federal Research Division, Library of Congress, 1989. As of January 6, 2010:
http://lcweb2.loc.gov/frd/cs/lbtoc.html

Adams, S., "*New York Times* Article Concerning VC Defections and Desertions, 18 December 1966," memorandum to the director of the Office of Current Intelligence, Central Intelligence Agency, December 19, 1966. As of January 7, 2010:
http://www.vietnam.ttu.edu/star/images/024/0240519020.pdf

Advameg, "Angola: A Country Study," *Encyclopedia of the Nations*, undated Web page. As of April 15, 2009:
http://www.country-data.com/frd/cs/aotoc.html

"Al-Qaida on the Fall of Baghdad, Guerilla Warfare," Middle East Media Research Institute, Special Dispatch Series, No. 493. As of December 1, 2006:
http://memri.org/bin/articles.cgi?Page=subjects&Area=jihad&ID=SP49303

Allnutt, Bruce C., *Marine Combined Action Capabilities: The Vietnam Experience*, McLean, Va.: Office of Naval Research, December 1969.

Alonso, Rogelio, "The Modernization in Irish Republican Thinking Toward the Utility of Violence," *Studies in Conflict and Terrorism*, Vol. 24, No 2, January 2001, pp. 131–144.

Anderson, David, *Histories of the Hanged: Britain's Dirty War in Kenya and the End of Empire*, London: Phoenix, 2005.

Arreguín-Toft, Ivan, *How the Weak Win Wars: A Theory of Asymmetric Conflict*, New York: Cambridge University Press, 2007.

Athas, Iqbal, "Safe House Raid: Heads Roll as Army Chief Cracks the Whip," *Sunday Times* (Sri Lanka), January 25, 2004a.

———, "Peace Talks: LTTE Not Likely to Respond Soon," *Sunday Times* (Sri Lanka), April 25, 2004b.

———, "Peace Process Bogged Down in More Questions," *Sunday Times* (Sri Lanka), May 2, 2004c.

Badeeb, Saeed M., *The Saudi-Egyptian Conflict over North Yemen, 1962–1970*, Boulder, Colo.: Westview Press, 1986.

Baker, James Addison, Lee Hamilton, and Lawrence S. Eagleburger, *The Iraq Study Group Report*, New York: Vintage Books, 2006. As of January 6, 2010: http://www.usip.org/programs/initiatives/iraq-study-group

Balakrishnan, H., "From Naxalbari to Nalgonda," *Hindu*, December 5, 2004. As of January 7, 2010: http://www.hindu.com/mag/2004/12/05/stories/2004120500470400.htm

Bamford, Bradley W. C., "The Role and Effectiveness of Intelligence in Northern Ireland," *Intelligence and National Security*, Vol. 20, No. 4, December 2005, pp. 581–607.

Beckett, I. F. W., *Modern Insurgencies and Counter-Insurgencies: Guerrillas and Their Opponents Since 1750*, London and New York: Routledge, 2001.

"Behind the Tamil Tigers," SBS Dateline (Australia), October 4, 2000.

Bell, J. Bowyer, "The Armed Struggle and Underground Intelligence: An Overview," *Studies in Conflict and Terrorism*, Vol. 17, No. 2, 1994, pp. 115–150.

———, *The Secret Army: The IRA*, revised 3rd ed., Dublin, Ireland: Poolbeg, 1998.

Bell, Stewart, "Underground to Canada," *National Post* (Canada), March 25, 2000a.

———, "Ripped Off, Abandoned and Broke," *National Post* (Canada), March 28, 2000b.

———, "For Sale in Toronto: Fake Roma Papers," *National Post* (Canada), March 29, 2000c.

———, "Money Trail: Financing War from Canada," *National Post* (Canada), June 3, 2000d.

Benramdane, Djamel, "Algeria Accepts the Unacceptable," *Le Monde Diplomatique*, March 1999. As of December 2, 2006: http://mondediplo.com/1999/03/03algeria

Bethell, Leslie, ed., *The Cambridge History of Latin America*, Vol. VII: *Latin America Since 1930, Mexico, Central America and the Caribbean*, New York: Cambridge University Press, 1990.

Bhattarai, Baburam, chair, Central Committee, United People's Front, Nepal, memorandum to the prime minister of Nepal, February 4, 1996.

Bickel, P. J., E. A. Hammel, and J. W. O'Connell, "Sex Bias in Graduate Admissions: Data from Berkeley," *Science*, Vol. 187, No. 4175, February 7, 1975, pp. 398–404.

Blaufarb, Douglas S., *The Counterinsurgency Era: U.S. Doctrine and Performance, 1950 to the Present*, New York: Free Press, 1977.

Bonner, Raymond, "Rebels in Sri Lanka Fight with Aid of Global Market in Light Arms," *New York Times*, March 7, 1998.

Bridgland, Fred, "Savimbi Death Opens Way for Angola to Leave Decades of Bloodshed Behind," *Scotsman Online*, February 25, 2002. As of January 6, 2010: http://news.scotsman.com/angola/Savimbi-death-opens-way-for.2305259.jp

Brookings Institute, "Iraq Index," last updated December 11, 2009. As of November 2009: http://www.brookings.edu/saban/iraq-index.aspx

Burns, Rupert, "Terrorism at the Beginning of the 21st Century," presentation, Institute of Defence and Strategic Studies Maritime Terrorism Conference, Singapore, May 20–21, 2004.

Byman, Daniel, "Passive Sponsors of Terrorism," *Survival*, Vol. 47, No. 4, Winter 2005, pp. 117–144. As of January 6, 2010: http://www.brookings.edu/articles/2005/winter_middleeast_byman.aspx

———, *Understanding Proto-Insurgencies: RAND Counterinsurgency Study— Paper 3*, Santa Monica, Calif.: RAND Corporation, OP-178-OSD, 2007. As of January 7, 2010: http://www.rand.org/pubs/occasional_papers/OP178/

Byman, Daniel, Peter Chalk, Bruce Hoffman, William Rosenau, and David Brannan, *Trends in Outside Support for Insurgent Movements*, Santa Monica, Calif.: RAND Corporation, MR-1405-OTI, 2001. As of January 6, 2010: http://www.rand.org/pubs/monograph_reports/MR1405/

Caldwell, C. E., *Small Wars: Their Principles and Practice*, 3rd ed., Lincoln, Neb.: University of Nebraska Press, 1906 (1996).

Cassidy, Robert M., *Counterinsurgency and the Global War on Terror: Military Culture and Irregular War*, Stanford, Calif.: Stanford University Press, 2008.

Central Intelligence Agency, *Guide to the Analysis of Insurgency*, 1986.

Chalk, Peter, *The Malay-Muslim Insurgency in Southern Thailand—Understanding the Conflict's Evolving Dynamic: RAND Counterinsurgency Study—Paper 5,* Santa Monica, Calif.: RAND Corporation, OP-198-OSD, 2008. As of January 12, 2010:
http://www.rand.org/pubs/occasional_papers/OP198/

Chalk, Peter, and Bruce Hoffman, *The Dynamics of Suicide Terrorism: Four Case Studies of Terrorist Movements,* Santa Monica, Calif.: RAND Corporation, 2005. Not available to the general public.

Charters, David A., "Intelligence and Psychological Operations in Northern Ireland," *RUSI Journal,* Vol. 122, No. 3, September 1977, pp. 22–27.

Chernick, Marc, "The FARC-EP: From Liberal Guerrillas to Marxist Rebels to Post–Cold War Insurgents," in Marianne Heiberg, Brendan O'Leary, and John Tirman, eds., *Terror, Insurgencies, and the State: Ending Protracted Conflicts,* Philadelphia, Pa.: University of Pennsylvania Press, 2007, pp. 52–82.

Cleen Foundation, "Graphical Representation of Crime Statistics 1994–2003," undated. As of January 12, 2010:
http://www.cleen.org/crime%20statistics%201994-2003_graphics.pdf

Clodfelter, Micheal, *Vietnam in Military Statistics: A History of the Indochina Wars, 1772–1991,* Jefferson, N.C.: McFarland and Company, 1995.

Clutterbuck, Richard L., *The Long, Long War: Counterinsurgency in Malaya and Vietnam,* New York: Praeger, 1966.

———, *Guerrillas and Terrorists,* London: Faber and Faber, 1977.

Coll, Steve, *Ghost Wars: The Secret History of the CIA, Afghanistan, and bin Laden, from the Soviet Invasion to September 10, 2001,* New York: Penguin Press, 2004.

Collier, Paul, "Rebellion as a Quasi-Criminal Activity," *Journal of Conflict Resolution,* Vol. 44, No. 6, December 2000, pp. 839–853.

———, *Wars, Guns, and Votes: Democracy in Dangerous Places,* New York: Harper, 2009.

Collins, Eamon, and Mick McGovern, *Killing Rage,* London: Granta Books, 1998.

Connable, Ben, "The Massacre That Wasn't," in G. J. David and T. R. McKeldin, eds., *Ideas as Weapons: Information and Perception in Modern Warfare,* Washington, D.C.: Potomac Books, 2009.

Coogan, Tim Pat, *The IRA: A History,* Niwot, Colo.: Roberts Rinehart, 1993.

Cordesman, Anthony H., and Abraham R. Wagner, *The Lessons of Modern War,* London: Mansell Publishing Limited, 1990.

Corfield, F. D., *Historical Survey of the Origins of the Growth of the Mau Mau,* London: H. M. Stationery Office, 1960.

Correspondence with senior Sri Lankan intelligence official (name withheld on request), November 2006.

Crile, George, *Charlie Wilson's War: The Extraordinary Story of the Largest Covert Operation in History*, New York: Atlantic Monthly Press, 2003.

Daly, Sara, "The Algerian Salafist Group for Call and Combat: A Dossier," *Terrorism Monitor*, Vol. 3, No. 5, March 11, 2005. As of January 6, 2010: http://www.jamestown.org/single/?no_cache=1&tx_ttnews[tt_news]=27670

Davis, Anthony, "Tracking Tigers in Phuket," *Asiaweek*, Vol. 29, No. 23, June 16, 2000. As of January 6, 2010: http://cgi.cnn.com/ASIANOW/asiaweek/magazine/2000/0616/nat.security.html

———, "Tamil Tiger Arms Intercepted," *Jane's Intelligence Review*, February 1, 2004.

———, "Tamil Tigers Seek to Rebuild Naval Force," *Jane's Intelligence Review*, March 2005.

Debray, Régis, *Revolution in the Revolution? Armed Struggle and Political Struggle in Latin America*, New York: Grove Press, Inc., 1967.

Degregori, Carlos Iván, "Harvesting Storms: Peasant *Rondas* and the Defeat of Sendero Luminoso in Ayacucho," in Steve J. Stern, ed., *Shining and Other Paths: War and Society in Peru, 1980–1995*, Durham N.C.: Duke University Press, 1998, pp. 128–158.

Al-Dhari, Hareth, "To Attack Innocent Civilians Is Not Jihad!" *Irak Müslüman Alimler Heyeti*, April 9, 2007. As of January 20, 2010: http://www.heyetnet.org/eng/meetings/ 198-al-dhari-to-attack-innocent-civilians-is-not-jihad.html

Dixit, J., "Indian Involvement in Sri Lanka and the Indo-Sri Lanka Agreement of 1987: A Retrospective Evaluation," in Kumar Rupesinghe, ed., *Negotiating Peace in Sri Lanka: Efforts, Failures and Lessons*, London: International Alert, 1998.

Dobbins, James, John G. McGinn, Keith Crane, Seth G. Jones, Rollie Lal, Andrew Rathmell, Rachel M. Swanger, and Anga R. Timilsina, *America's Role in Nation-Building: From Germany to Iraq*, Santa Monica, Calif.: RAND Corporation, MR-1753-RC, 2003. As of January 7, 2010: http://www.rand.org/pubs/monograph_reports/MR1753/

Drake, C. J. M., "The Provisional IRA: A Case Study," *Terrorism and Political Violence*, Vol. 3, No. 2, Summer 1991, pp. 43–60.

Edwards, David B., *Before Taliban: Genealogies of the Afghan Jihad*, Berkeley, Calif.: University of California Press, 2002.

Elkins, Caroline, *Imperial Reckoning: The Untold Story of Britain's Gulag in Kenya*, New York: Henry Holt and Company, 2005.

Ellison, Graham, and Jim Smyth, *The Crowned Harp: Policing Northern Ireland*, London: Pluto Press, 2000.

Email correspondence between author and Sri Lankan intelligence official (name withheld on request), May 2005.

Ethirajan, Anbarasan, "Fierce Clashes in North Sri Lanka," *BBC*, March 30, 2009. As of January 12, 2010:
http://news.bbc.co.uk/2/hi/south_asia/7971389.stm

Europa World Year Book, London: Europa Publications, 1998.

Fahoum, Keely M., and Jon Width, "Marketing Terror: Effects of Anti-Messaging on GSPC Recruitment," *Strategic Insights*, Vol. 5, No. 8, November 2006. As of January 6, 2010:
http://www.au.af.mil/au/awc/awcgate/nps/fahoum_nov06.pdf

Fall, Bernard B., *Street Without Joy*, 4th ed., Harrisburg, Pa.: Stackpole Company, 1964.

Fearon, James D., and David D. Laitin, "Ethnicity, Insurgency, and Civil War," *American Political Science Review*, Vol. 97, No. 1, February 2003a, pp. 75–90.

———, "Additional Tables for 'Ethnicity, Insurgency, and Civil War,'" Stanford University, February 6, 2003b. As of January 6, 2010:
http://www.stanford.edu/group/ethnic/workingpapers/addtabs.pdf

Flynn, Michael T., Matt Pottinger, and Paul Batchelor, *Fixing Intel: A Blueprint for Making Intelligence Relevant in Afghanistan*, Washington, D.C.: Center for a New American Security, 2010. As of January 20, 2010:
http://www.cnas.org/files/documents/publications/
AfghanIntel_Flynn_Jan2010_code507_voices.pdf

FM 3-24—*see* U.S. Department of the Army and U.S. Marine Corps (2006).

Forero, Juan, "FARC Dissidents Assist Colombia: Jailed Rebels Share Inside Information," *Washington Post*, August 2, 2008. As of January 6, 2010:
http://www.washingtonpost.com/wp-dyn/content/article/2008/08/01/
AR2008080103117.html

Galula, David, *Counterinsurgency Warfare: Theory and Practice*, Westport, Conn.: Praeger Security International, 1964 (2006).

GAO—*see* U.S. Government Accountability Office.

Geraghty, Tony, *The Irish War: The Hidden Conflict Between the IRA and British Intelligence*, Baltimore, Md.: Johns Hopkins University Press, 2000.

Gilmour, David, *Lebanon: The Fractured Country*, New York: St. Martin's Press, 1983.

Giustozzi, Antonio, *Afghanistan: Transition Without End*, London: Crisis States Research Centre, working paper 40, series 2, September 2008. As of January 6, 2010:
http://www.crisisstates.com/Publications/wp/WP40.2.htm

Gladwell, Malcolm, *The Tipping Point: How Little Things Can Make a Big Difference*, New York: Little, Brown, 2000.

Glover, J. M., *Northern Ireland: Future Terrorist Trends*, London: Ministry of Defence, British Government, 1978.

Gourevitch, Philip, "Tides of War," *New Yorker*, August 1, 2005. As of January 6, 2010:
http://www.newyorker.com/archive/2005/08/01/050801fa_fact1

Grant, Audra, "The Algerian 2005 Amnesty: The Path to Peace?" *Terrorism Monitor*, Vol. 3, No. 22, November 17, 2005. As of January 6, 2010:
http://www.jamestown.org/single/?no_cache=1&tx_ttnews[tt_news]=614

Grau, Lester W., ed., *The Bear Went Over the Mountain: Soviet Combat Tactics in Afghanistan*, Washington, D.C.: National Defense University, December 1995. As of January 6, 2010:
http://www.ndu.edu/inss/books/Books%20-%201996/
Bear%20Went%20Over%20Mountain%20-%20Aug%2096/BrOrMn.pdf

Grynkewich, Alex, and Chris Reifel, "Modeling Jihad: A System Dynamics Model of the Salafist Group for Preaching and Combat Financial Subsystem," *Strategic Insights*, Vol. 5, No. 8, November 2006. As of January 6, 2010:
http://www.nps.edu/Academics/centers/ccc/publications/OnlineJournal/2006/
Nov/grynkewichNov06.html

Guelke, Adrian, "Civil Society and the Northern Irish Peace Process," *Voluntas: International Journal of Voluntary and Nonprofit Organizations*, Vol. 14, No. 1, March 2003, pp. 61–78.

Guevara, Ernesto "Che," *Guerrilla Warfare*, Thousand Oaks, Calif.: BN Publishing, 1969 (2008).

Guillén, Abraham, *Philosophy of the Urban Guerrilla: The Revolutionary Writings of Abraham Guillén*, New York: Morrow, 1973.

Gunaratna, Rohan, *War and Peace in Sri Lanka, with a Post-Accord Report from Jaffna*, Colombo: Institute of Fundamental Studies, Sri Lanka, 1987.

———, *International and Regional Security Implications of the Sri Lankan Tamil Insurgency*, Sri Lanka: Alumni Association of the Bandaranaike Centre for International Studies, 1997a.

———, "Illicit Transfer of Conventional Weapons: The Role of State and Non-State Actors in South Asia," paper presented, Third Inter-Sessional Workshop of the Panel of Governmental Experts on Small Arms, May 22–23, 1997b.

———, *Sri Lanka's Ethnic Crisis and National Security*, Colombo: South Asia Network on Conflict Research, 1998.

———, "LTTE Organization and Operations in Canada," unpublished document supplied to author, November 2000.

———, "Maritime Terrorism: Future Threats and Responses," paper delivered, International Research Group on Political Violence, United States Institute of Peace, Washington D.C., April 15, 2001a.

———, "Intelligence Failures Exposed by Tamil Tiger Airport Attack," *Jane's Intelligence Review*, September 1, 2001b.

Gurr, Ted Robert, "Persistence and Change in Political Systems, 1800–1971," *American Political Science Review*, Vol. 68, No. 4, December 1974, pp. 1482–1504.

Hamill, Desmond, *Pig in the Middle: The Army in Northern Ireland, 1969–1984*, London: Methuen, 1985.

Hammes, Thomas X., *The Sling and the Stone: On War in the 21st Century*, St. Paul, Minn.: Zenith, 2006.

Hauser, Christine, "Scores Are Killed in Attack on Sri Lankan Convoy," *New York Times*, October 16, 2006. As of January 6, 2010: http://www.nytimes.com/2006/10/16/world/asia/17lankacnd.html

Heiberg, Marianne, Brendan O'Leary, and John Tirman, eds., *Terror, Insurgency, and the State: Ending Protracted Conflicts*, Philadelphia, Pa.: University of Pennsylvania Press, 2007.

Hoffman, Bruce, and Jennifer Taw, *A Strategic Framework for Countering Terrorism and Insurgency*, Santa Monica, Calif.: RAND Corporation, N-3506-DOS, 1992. As of January 6, 2010: http://www.rand.org/pubs/notes/N3506/

Hoffman, Frank G., *Hybrid Threats: Reconceptualizing the Evolving Character of Modern Conflict*, Institute for National Strategic Studies, Strategic Forum 240, April 2009.

Holland, Jack, and Susan Phoenix, *Phoenix: Policing the Shadows*, London: Hodder and Stoughton, 1996.

Horgan, John, and Max Taylor, "Playing the 'Green Card'—Financing the Provisional IRA: Part 1," *Terrorism and Political Violence*, Vol. 11, No. 2, Summer 1999, pp. 1–38.

Hosmer, Stephen T., *Counterinsurgency: A Symposium, April 16–20, 1962*, Santa Monica, Calif.: RAND Corporation, R-412-1-ARPA/RC, 1963 (2006). As of January 6, 2010: http://www.rand.org/pubs/reports/R412-1/

"Hostages Held for More Than Five Years Rescued in Colombia," Cable News Network, July 3, 2008. Transcript, as of February 11, 2010: http://transcripts.cnn.com/TRANSCRIPTS/0807/02/acd.02.html

Hughes, Geraint, "A 'Model Campaign' Reappraised: The Counter-Insurgency War in Dhofar, Oman, 1965–1975," *Journal of Strategic Studies*, Vol. 32, No. 2, April 2009, pp. 271–305.

Hughes, Joanne, and Caitlin Donnelly, "Attitudes to Community Relations in Northern Ireland: Signs of Optimism in the Post Cease-Fire Period?" *Terrorism and Political Violence*, Vol. 16, No. 3, August 2004, pp. 567–592.

Hunt, Emily, "Al-Qaeda's North African Franchise: The GSPC Regional Threat," *PolicyWatch* 1034, Washington Institute for Near East Policy, September 28, 2005. As of November 21, 2006: http://www.washingtoninstitute.org/templateC05.php?CID=2379

International Crisis Group, *The Civil Concord: A Peace Initiative Wasted*, Africa Report No. 31, July 9, 2001. As of January 6, 2010: http://www.crisisgroup.org/home/index.cfm?l=1&id=1419

———, *Islam, Violence and Reform in Algeria: Turning the Page*, Islamism in North Africa III, Middle East Report No. 29, July 30, 2004. As of January 6, 2010: http://www.crisisgroup.org/home/index.cfm?id=2884&l=1

International Institute for Strategic Studies, "Sri Lanka's Peace Process in Jeopardy," *Strategic Comments*, Vol. 10, No. 3, April 2004, pp. 1–2.

Interview with former law-enforcement official (name withheld on request), Northern Ireland, May 2005.

Interview with Gérard Chaliand, director, Centre Européen d'Etude des Conflits, Jaffna, May 1999.

Interview with law-enforcement official (name withheld on request), Northern Ireland, May 2005.

Interview with Thailand Internal Intelligence Bureau official (name withheld on request), Bangkok, April 2005.

Interview with Western security analyst, Bangkok, September 2006.

Interviews with Internal Intelligence Bureau officials (names withheld on request), Colombo, May 2004.

Interviews with senior Western diplomatic officials, Colombo, May 2004.

Interviews with Sri Lankan and Western officials, May 2004.

Interviews with Sri Lankan Armed Forces officials (names withheld on request), Colombo, May 2004.

Interviews with Sri Lankan commentators and intelligence officials (names withheld on request), Bangkok, November 2005.

Interviews with Sri Lankan diplomatic officials (names withheld on request), Bangkok, December 2000.

Interviews with Sri Lankan government officials (names withheld on request), Canberra, Australia, December 2000.

Interviews with Sri Lankan intelligence and military officials (names withheld on request), Bangkok and Colombo, May 2004 and April 2005.

Interviews with Sri Lankan observers and security officials (names withheld on request), Ottawa, Canada, and Bangkok, Thailand, November–December 2000.

Interviews with Sri Lankan officials and former Liberation Tigers of Tamil Eelam cadres (names withheld on request), Colombo and Jaffna, May 1999.

Jackson, Brian A., "Counterinsurgency Intelligence in a 'Long War': The British Experience in Northern Ireland," *Military Review*, January–February 2007, pp. 74–85. Reprint, as of January 6, 2010: http://www.rand.org/pubs/reprints/RP1247/

Jalali, Ali Ahmad, and Lester W. Grau, *The Other Side of the Mountain: Mujahideen Tactics in the Soviet-Afghan War*, Quantico, Va.: U.S. Marine Corps Studies and Analysis Division, 1999. As of January 6, 2010: http://purl.access.gpo.gov/GPO/LPS72248

Jayasekera, Bandula, "Prabhakaran Smuggled in 11 Arms Shiploads During Truce: Karuna," *Island* (Sri Lanka), April 9, 2004.

Jayasinghe, Amal, "Tiger Bombers Primed for a Repeat," *Australian*, February 8, 1996a.

———, "Two Tamil Tiger Bombs Kill 70 in Colombo Commuter Train," *Guardian* (UK), June 25, 1996b.

Joes, Anthony James, *Resisting Rebellion: The History and Politics of Counterinsurgency*, Lexington, Ky.: University of Kentucky Press, 2004.

Jones, Seth G., Jeremy M. Wilson, Andrew Rathmell, and K. Jack Riley, *Establishing Law and Order After Conflict*, Santa Monica, Calif.: RAND Corporation, MG-374-RC, 2005. As of January 6, 2010: http://www.rand.org/pubs/monographs/MG374/

Joshi, Charu Lata, "The Body Trade," *Far Eastern Economic Review*, Vol. 163, October 26, 2000, pp. 100–104.

Joshi, Manoj, "On the Razor's Edge: The Liberation Tigers of Tamil Eelam," *Studies on Conflict and Terrorism*, Vol. 19, No. 1, January–March 1996, pp. 19–42.

JP 3-24—*see* U.S. Department of the Army and U.S. Marine Corps (2009).

Judge, Paramjit S., *Insurrection to Agitation: The Naxalite Movement in Punjab*, Bombay: Popular Prakashan, 1992.

Kalyvas, Stathis N., *The Logic of Violence in Civil War*, New York: Cambridge University Press, 2006.

Kamalendran, Chris, "Inside the Karuna Fortress," *Sunday Times* (Sri Lanka), March 14, 2004. As of January 6, 2010:
http://sundaytimes.lk/040314/

Kane, Tim, *Global U.S. Troop Deployment, 1950–2005*, Washington, D.C., Heritage Foundation, May 24, 2006. As of January 12, 2010:
http://www.heritage.org/Research/NationalSecurity/cda06-02.cfm

Keen, David, *The Economic Functions of Violence in Civil Wars*, Oxford, UK: International Institute for Strategic Studies, Adelphi paper 320, 1997.

Kenyatta, Jomo, *Facing Mount Kenya: The Tribal Life of the Gikuyu*, New York: Vintage Books, 1965.

Kilcullen, David, *Countering Global Insurgency: A Strategy for the War on Terrorism*, Canberra and Washington, D.C., 2004.

———, *Twenty-Eight Articles: Fundamentals of Company-Level Counterinsurgency*, Washington, D.C., March 2006.

———, *The Accidental Guerrilla: Fighting Small Wars in the Midst of a Big One*, Oxford, UK: Oxford University Press, 2009.

Kilcullen, David, and Andrew McDonald Exum, "Death from Above, Outrage Down Below," *New York Times*, May 16, 2009. As of January 6, 2010:
http://www.nytimes.com/2009/05/17/opinion/17exum.html

King, Charles, *Ending Civil Wars*, Oxford, UK: International Institute for Strategic Studies, Adelphi paper 308, 1997.

Kitson, Frank, *Bunch of Five*, London: Faber, 1977.

Klinghoffer, Arthur Jay, *The Angolan War: A Study in Soviet Policy in the Third World*, Boulder, Colo.: Westview Press, 1980.

Koch, Jeanette A., *The Chieu Hoi Program in South Vietnam, 1963–1971*, Santa Monica, Calif.: RAND Corporation, R-1172-ARPA, 1973. As of January 7, 2010:
http://www.rand.org/pubs/reports/R1172/

Kohl, James, and John Litt, *Urban Guerrilla Warfare in Latin America*, Cambridge, Mass.: MIT Press, 1974.

Konogo, Tabitha M., *Squatters and the Roots of the Mau Mau, 1905–63*, Athens, Ohio: Ohio University Press, 1987.

Kraul, Chris, "The Shine Is Off Colombia's FARC," *Los Angeles Times*, January 19, 2009.

Krause, Peter J. P., "Troop Levels in Stability Operations: What We Don't Know," MIT Center for International Studies, February 2007.

Krayem, Hassam, "The Lebanese Civil War and the Taif Agreement," American University of Beirut, undated. As of January 6, 2010: http://ddc.aub.edu.lb/projects/pspa/conflict-resolution.html

Krepinevich, Andrew J., "Are We Winning in Iraq?" testimony before U.S. House of Representatives Committee on Armed Services, March 17, 2005a.

———, "How to Win in Iraq," *Foreign Affairs*, September–October 2005b.

Kulikov, Sergey A., "Insurgent Groups in Chechnya," Robert R. Love, trans., *Military Review*, November–December 2003, pp. 21–30. As of January 12, 2010: http://usacac.leavenworth.army.mil/CAC/milreview/download/english/NovDec03/kulikov.pdf

Lanning, Michael Lee, and Dan Cragg, *Inside the VC and the NVA: The Real Story of North Vietnam's Armed Forces*, New York: Fawcett Columbine, 1992.

Lapper, Richard, "Uribe Ascendant: Defeats for the FARC Mark a Shift of Power in Latin America," *Financial Times Online*, July 6, 2008.

Laqueur, Walter, *Guerrilla: A Historical and Critical Study*, Boston, Mass.: Little, Brown, 1976.

Lawrence, T. E., *Seven Pillars of Wisdom: A Triumph*, London: Cape, 1990.

Layachi, Azzedine, and John P. Entelis, "Democratic and Popular Republic of Algeria," in David E. Long and Bernard Reich, eds., *The Government and Politics of the Middle East and North Africa*, 4th ed., Oxford, UK: Westview Press, 2002, pp. 423–460.

Leakey, L. S. B., *Defeating Mau Mau*, London: Methuen, 1954.

Li, Xiaobing, *A History of the Modern Chinese Army*, Lexington, Ky.: University Press of Kentucky, 2007.

Liakhovskii, Aleksandr Antonovich, *Inside the Soviet Invasion of Afghanistan and the Seizure of Kabul, December 1979*, Washington, D.C.: Woodrow Wilson International Center for Scholars, Cold War International History Project, working paper 51, January 2007.

Lintner, Bertil, "The Phuket Connection," *Week*, April 30, 2000.

Malik, Habib C., *Between Damascus and Jerusalem: Lebanon and Middle East Peace*, Washington, D.C.: Washington Institute for Near East Policy, policy paper 45, 2000.

Malkasian, Carter, "Did the Coalition Need More Forces in Iraq? Evidence from Al Anbar," *Joint Forces Quarterly*, Issue 46, 3rd Quarter 2007, pp. 120–126. As of January 6, 2010: http://www.ndu.edu/inss/Press/jfq_pages/editions/i46/19.pdf

Mallie, Eamonn, and David McKittrick, *The Fight for Peace: The Secret Story Behind the Irish Peace Process*, London: Heinemann, 1996.

Maloba, Wunyabari O., *Mau Mau and Kenya: An Analysis of a Peasant Revolt*, Bloomington, Ind.: Indiana University Press, 1993.

Mao Tse-tung, *Selected Works*, New York: International Publishers, 1954–1962.

———, *Selected Works*, Vol. I: *On the Tactics of Fighting Japanese Imperialism*, New York: International Publishers, 1956.

———, *On Guerrilla Warfare*, Samuel B. Griffith, trans., Urbana, Ill.: University of Illinois Press, 1961 (2000).

———, *Basic Tactics*, Stuart R. Schram, trans., New York: Praeger, 1966.

Marcum, John A., "Lessons of Angola," *Foreign Affairs*, Vol. 54, No. 3, April 1976.

———, *The Angolan Revolution*, Vol. II: *Exile Politics and Guerrilla Warfare (1962–1976)*, Cambridge, Mass.: MIT Press, 1978.

Marighella, Carlos, *Minimanual of the Urban Guerrilla*, St. Petersburg, Fla.: Red and Black Publishers, 1969 (2008).

———, *For the Liberation of Brazil*, Harmondsworth, UK: Penguin Books, 1971.

Markel, W., "Draining the Swamp: The British Strategy of Population Control," *Parameters: U.S. Army War College Quarterly*, Carlisle, Pa.: U.S. Army War College, Vol. XXXVI, No. 1, Spring 2006.

Martínez, Luis, *La guerre civile en Algérie, 1990–1998* [The Algerian Civil War, 1990–1998], Paris: Karthala, 1998.

"Masked Gunmen Kill Jaffna Mayor," *Daily News* (Sri Lanka), July 28, 1975.

Mason, Robert Scott, "The Opposing Forces in the Lebanese Civil War," in Thomas Collelo and Harvey Henry Smith, eds., *Lebanon: A Country Study*, 3rd ed., Washington, D.C., Federal Research Division, Library of Congress, 1989a. As of January 6, 2010:
http://lcweb2.loc.gov/frd/cs/lbtoc.html

———, "The Riyadh Conference and the Arab Deterrent Force," in Thomas Collelo and Harvey Henry Smith, eds., *Lebanon: A Country Study*, 3rd ed., Washington, D.C., Federal Research Division, Library of Congress, 1989b. As of January 6, 2010:
http://lcweb2.loc.gov/frd/cs/lbtoc.html

McConnell, John Alexander, *The British in Kenya (1952–1960): Analysis of a Successful Counterinsurgency Campaign*, Monterey, Calif.: Naval Postgraduate School, 2005. As of January 6, 2010:
http://handle.dtic.mil/100.2/ADA435532

McCormick, Gordon, *From the Sierra to the Cities: The Urban Campaign of the Shining Path*, Santa Monica, Calif.: RAND Corporation, R-4150-USDP, 1992. As of January 12, 2010:
http://www.rand.org/pubs/reports/R4150/

McCormick, Gordon H., Steven B. Horton, and Lauren A. Harrison, "Things Fall Apart: The Endgame Dynamics of Internal Wars," paper presented to the RAND Corporation Insurgency Board, August 29, 2006.

McCuen, John J., *The Art of Counter-Revolutionary War: The Strategy of Counter-Insurgency*, London: Faber, 1966.

———, "Hybrid Wars," *Military Review*, March–April 2008, pp. 107–113. As of January 6, 2010:
http://www.au.af.mil/au/awc/awcgate/milreview/mccuen08marapr.pdf

McGarry, John, "How the IRA Was Tamed," *Globe and Mail*, August 2, 1005, p. A17.

McGartland, Martin, *Fifty Dead Men Walking: The Heroic True Story of a British Secret Agent Inside the IRA*, London: Blake, 1997.

McGovern, Mark, "'The Old Days Are Over': Irish Republicanism, the Peace Process and the Discourse of Equality," *Terrorism and Political Violence*, Vol. 16, No. 3, Autumn 2004, pp. 622–645.

McGrath, John J., *Boots on the Ground: Troop Density in Contingency Operations*, Fort Leavenworth, Kan.: Combat Studies Institute Press, Global War on Terrorism occasional paper 16, 2006. As of February 11, 2010:
http://cgsc.leavenworth.army.mil/carl/download/csipubs/mcgrath_boots.pdf

McKinley, Michael, "The International Dimensions of Irish Terrorism," in Yonah Alexander and Alan O'Day, eds., *Terrorism in Ireland*, London: Croom Helm, 1984.

Merom, Gil, *How Democracies Lose Small Wars: State, Society, and the Failures of France in Algeria, Israel in Lebanon, and the United States in Vietnam*, New York: Cambridge University Press, 2003.

Miller, William H., "Insurgency Theory and the Conflict in Algeria: A Theoretical Analysis," *Terrorism and Political Violence*, Vol. 12, No. 1, Spring 2000, pp. 60–78.

Mortimer, Robert, "Bouteflika and Algeria's Path from Revolt to Reconciliation," *Current History*, Vol. 99, No. 633, January 2000.

Murray, J. C., "The Anti-Bandit War," in Thomas Nicholls Greene, *The Guerrilla, And How to Fight Him: Selections from the Marine Corps Gazette*, Quantico, Va.: Marine Corps Association, 1962.

Mydans, Seth, "Outcome of Sri Lanka's Long War May Hang on Fate of Insurgent Leader," *New York Times*, March 31, 2009. As of April 1, 2009:
http://www.nytimes.com/2009/04/01/world/asia/01lanka.html

Nagl, John A., *Learning to Eat Soup with a Knife: Counterinsurgency Lessons from Malaya and Vietnam*, Chicago, Ill.: University of Chicago Press, 2005.

Neumann, Peter R., "From Revolution to Devolution: Is the IRA Still a Threat to Peace in Northern Ireland?" *Journal of Contemporary European Studies*, Vol. 13, No. 1, April 2005a, pp. 79–92.

———, "The Bullet and the Ballot Box: The Case of the IRA," *Journal of Strategic Studies*, Vol. 28, No. 6, December 2005b, pp. 941–975.

Nicol, John, "Passports for Sail," *McLeans*, April 3, 2000.

O'Callaghan, Sean, *The Informer*, London: Corgi, 1999.

O'Duffy, Brendan, "Majoritarianism, Self-Determination, and Military-to-Political Transition in Sri Lanka," in Marianne Heiberg, Brendan O'Leary, and John Tirman, eds., *Terror, Insurgency, and the State: Ending Protracted Conflicts*, Philadelphia, Pa.: University of Pennsylvania Press, 2007, pp. 257–288.

Oliker, Olga, *Russia's Chechen Wars 1994–2000: Lessons from Urban Combat*, Santa Monica, Calif.: RAND Corporation, MR-1289-A, 2001. As of January 12, 2010:
http://www.rand.org/pubs/monograph_reports/MR1289/

Olson, Eric T., commanding general, U.S. Special Operations Command, "SOCOM Posture Statement to Congress," 2009.

O'Neill, Bard E., *Insurgency and Terrorism: Inside Modern Revolutionary Warfare*, Washington, D.C.: Brassey's (US), 1990.

OSLO Conference on "Road Map to Peace in Sri Lanka," Oslo Kommune, Norway: World Alliance for Peace in Sri Lanka, 2005.

Pact of Caracas, July 20, 1958.

Page, Michael von Tagen, and M. L. R. Smith, "War by Other Means: The Problem of Political Control in Irish Republican Strategy," *Armed Forces and Society*, Vol. 27, No. 1, Fall 2000, pp. 79–104.

Palmer, David Scott, ed., *The Shining Path of Peru*, 2nd ed., New York: St. Martin's Press, 1994.

Paul, James, and Martin Spirit, "British Military Strategy in Kenya," undated. As of November 13, 2006:
http://www.britains-smallwars.com/kenya/Strategy.html

"Peace Process Bogged Down in More Questions," *Sunday Times* (Sri Lanka), May 2, 2004.

"Peace Talks: LTTE Not Likely to Respond Soon," *Sunday Times* (Sri Lanka), April 25, 2004.

Petraeus, GEN David, "Fox News Sunday with Chris Wallace," *Fox News Sunday*, June 17, 2007. As of January 20, 2010:
http://www.foxnews.com/story/0,2933,283553,00.html

Ponnambalam, Kumar, "The Only Possible Solution to the Tamil National Problem," *Weekend Express*, November 7–8, 1998.

Poroskov, Nikolai, "The USSR Was Actively Involved in the War in Vietnam 30 Years Ago," *Pravda*, April 30, 2005. As of January 7, 2010:
http://english.pravda.ru/main/18/90/363/15388_vietnam.html

Potgieter, Jakkie, "'Taking Aid from the Devil Himself': UNITA's Support Structures," in Jakkie Cilliers and Christian Dietrich, eds., *Angola's War Economy: The Role of Oil and Diamonds*, Pretoria: Institute for Security Studies, 2000, pp. 255–273. As of January 7, 2010:
http://www.iss.co.za/PUBS/Books/Angola/13Potgieter.pdf

Public Broadcasting Service, "Thirty Years of America's Drug War: A Chronology," *Frontline*, undated. As of April 1, 2009:
http://www.pbs.org/wgbh/pages/frontline/shows/drugs/cron/

Quinlivan, James T., "Force Requirements in Stability Operations," *Parameters*, Vol. 25, Winter 1995, pp. 59–69. As of January 7, 2010:
http://www.usamhi.army.mil/USAWC/Parameters/1995/quinliv.htm

———, "Burden of Victory: The Painful Arithmetic of Stability Operations," *RAND Review*, Vol. 27, No. 2, Summer 2003, pp. 28–29. As of January 7, 2010:
http://www.rand.org/publications/randreview/issues/summer2003/burden.html

Rabasa, Angel, and Peter Chalk, *Colombian Labyrinth: The Synergy of Drugs and Insurgency and Its Implications for Regional Stability*, Santa Monica, Calif.: RAND Corporation, MR-1339-AF, 2001. As of January 7, 2010:
http://www.rand.org/pubs/monograph_reports/MR1339/

Rabasa, Angel, Lesley Anne Warner, Peter Chalk, Ivan Khilko, and Paraag Shukla, *Money in the Bank—Lessons Learned from Past Counterinsurgency (COIN) Operations: RAND Counterinsurgency Study—Paper 4*, Santa Monica, Calif.: RAND Corporation, OP-185-OSD, 2007. As of January 7, 2010:
http://www.rand.org/pubs/occasional_papers/OP185/

Rabinovich, Itamar, *The War for Lebanon, 1970–1985*, Ithaca, N.Y.: Cornell University Press, 1984.

Raman, B., "Split in LTTE: The Clash of the Tamil Warlords," South Asia Policy Institute, topical paper 942, March 8, 2004. As of January 7, 2010:
http://www.saag.org/common/uploaded_files/paper942.html

———, "Sri Lanka: A Heavy Price for Over-Confidence," South Asia Analysis Group International Terrorism Monitor paper 139, October 14, 2006. As of January 7, 2010:
http://www.southasiaanalysis.org/\papers20\paper1989.html

"Rebel Commander Willing to Meet LTTE Top Leadership," Associated Press, March 11, 2004.

"Rebels Apologize for Rajiv Gandhi's Assassination," *New York Times*, June 28, 2006. As of January 7, 2010:
http://www.nytimes.com/2006/06/28/world/asia/28briefs-002.html

Record, Jeffrey, *Beating Goliath: Why Insurgencies Win*, Washington, D.C.: Potomac Books, 2007.

Richards, A., "Terrorist Groups and Political Fronts: The IRA, Sinn Fein, the Peace Process and Democracy," *Terrorism and Political Violence*, Vol. 13, No. 4, Winter 2001, pp. 72–89.

Roberts, John, "The Economy," in Thomas Collelo and Harvey Henry Smith, eds., *Lebanon: A Country Study*, 3rd ed., Washington, D.C., Federal Research Division, Library of Congress, 1989. As of January 6, 2010:
http://lcweb2.loc.gov/frd/cs/lbtoc.html

Rosenau, William, "Waging the 'War of Ideas,'" in David G. Kamien, ed., *The McGraw-Hill Homeland Security Handbook*, New York: McGraw-Hill, 2006, pp. 1131–1148.

Ross, James, "When Ceasefires Fail," *Foreign Policy in Focus*, September 15, 2006. As of January 7, 2010:
http://www.fpif.org/articles/when_ceasefires_fail

Rubin, Alissa J., and Damien Cave, "In a Force for Iraqi Calm, Seeds of Conflict," *New York Times*, December 23, 2007, p. A1. As of January 7, 2010:
http://www.nytimes.com/2007/12/23/world/middleeast/23awakening.html

Rue, John E., *Mao Tse-Tung in Opposition, 1927–1935*, Stanford, Calif.: Stanford University Press, 1966.

"Safe Haven—Facts and Prospects," Foreign and Commonwealth Office, Diplomatic Report 18/72, British Public Records Office, 33-1593/4, December 21, 1971.

Sapone, Montgomery, "Ceasefire: The Impact of Republican Political Culture on the Ceasefire Process in Northern Ireland," *Peace and Conflict Studies*, Vol. 7, No. 1, May 2000.

Sarma, Kiran, "Informers and the Battle Against Republican Terrorism: A Review of 30 Years of Conflict," *Police Practice and Research*, Vol. 6, No. 2, May 2005, pp. 165–180.

Schaffer, Teresita C., and Aneesh Deshpande, "Can Sri Lanka Turn Away from War?" *South Asia Monitor*, No. 96, July 5, 2006. As of January 6, 2010:
http://csis.org/publication/
south-asia-monitor-can-sri-lanka-turn-away-war-july-5-2006

Schwarz, Benjamin, *American Counterinsurgency Doctrine and El Salvador: The Frustrations of Reform and the Illusions of Nation Building*, Santa Monica, Calif.: RAND Corporation, R-4042-USDP, 1991. As of January 7, 2010: http://www.rand.org/pubs/reports/R4042/

Senanayake, Shimali, "Sri Lanka: Fierce Fighting Rages; Sweden Pulls Out," *New York Times*, August 2, 2006. As of January 7, 2010: http://query.nytimes.com/gst/fullpage.html?res=9507E7DA103FF931A3575BC0A9609C8B63

Sengupta, Samini, "Hard-Liner Is Elected Sri Lankan President," *International Herald Tribune*, November 19–20, 2005.

Shrader, Charles R., *The Withered Vine: Logistics and the Communist Insurgency in Greece, 1945–1949*, Westport, Conn.: Praeger, 1999.

Sinno, Abdulkader H., *Organizations at War in Afghanistan and Beyond*, Ithaca, N.Y.: Cornell University Press, 2008.

Smith, Chris, "South Asia's Enduring War," in Robert I. Rotberg, ed., *Creating Peace in Sri Lanka: Civil War and Reconciliation*, Washington, D.C.: Brookings Institution Press, 1999, pp. 17–40.

———, "Tamil Tigers Face Tough Choices in Wake of Tsunami," *Jane's Intelligence Review*, March 1, 2005.

———, "Rebel Attacks Intensify as Sri Lanka Slides Towards Civil War," *Jane's Intelligence Review*, March 1, 2006.

Smith, M. L. R., *Fighting for Ireland? The Military Strategy of the Irish Republican Movement*, London: Routledge, 1997.

"Sri Lanka Arrests 12 from Tamil Minority in Killing of Official," *New York Times*, August 15, 2005. As of January 7, 2010: http://query.nytimes.com/gst/fullpage.html?res=9D03E4DE123EF936A2575BC0A9639C8B63

"State of Emergency Prolonged," *New York Times*, July 7, 2006. As of January 7, 2010: http://query.nytimes.com/gst/fullpage.html?res=9803E0DF1030F934A35754C0A9609C8B63

Stern, Steve J., ed., *Shining and Other Paths: War and Society in Peru, 1980–1995*, Durham, N.C.: Duke University Press, 1998.

Stevenson, Jonathan, "Northern Ireland: Treating Terrorists as Statesmen," *Foreign Policy*, Vol. 105, Winter 1996–1997, pp. 125–140.

———, "Irreversible Peace in Northern Ireland?" *Survival*, Vol. 42, No. 3, January 2000, pp. 5–26.

Taber, Robert, *The War of the Flea: A Study of Guerrilla Warfare Theory and Practise*, New York: L. Stuart, 1965.

"Tamil Tigers Agree to Talks," *SBS World News*, August 19, 2005.

Tanner, Stephen, *Afghanistan: A Military History from Alexander the Great to the Fall of the Taliban*, New York: Da Capo Press, 2002.

Taylor, Peter, *Brits: The War Against the IRA*, London: Bloomsbury, 2001.

"Tet Offensive: Turning Point in Vietnam War," *New York Times*, January 31, 1988. As of January 7, 2010:
http://www.nytimes.com/1988/01/31/world/
tet-offensive-turning-point-in-vietnam-war.html

Thackrah, John Richard, *Encyclopedia of Terrorism and Political Violence*, London: Routledge and Kegan Paul, 1987.

Thomas, Raju G. C., "Secessionist Movements in South Asia," *Survival*, Vol. 36, No. 2, Summer 1994, pp. 92–114.

Thompson, Leroy, *The Counter-Insurgency Manual: Tactics of the Anti-Guerrilla Professionals*, London: Greenhill, 2002.

Thucydides, *History of the Peloponnesian War*, R. Warner, trans., Harmondsworth, UK: Penguin, 1972.

"The Tiger Shipping Empire," *Island* (Sri Lanka), March 26, 2000.

"Tigers Lashed Over Voter Intimidation," *South China Morning Post*, November 22, 2005.

Tilakaratna, Berbard, "The Sri Lanka Government and Peace Efforts Up to the Indo-Sri Lanka Accord: Lessons and Experiences," in Kumar Rupesinghe, ed., *Negotiating Peace in Sri Lanka: Efforts, Failures and Lessons*, London: International Alert, 1998.

Tower, John G., "Congress Versus the President: The Formulation and Implementation of American Foreign Policy," *Foreign Affairs*, Vol. 60, Number 2, Winter 1981–1982, pp. 237–238.

Tucker, H. H., *The IRA in Eire*, Foreign and Commonwealth Office, Republic of Ireland, registered files (WL series) 1972–1974.

UK Ministry of Defence, *Army Field Manual*, Vol. 1, 2001.

Ulph, Stephen, "Schism and Collapse of Morale in Algeria's GSPC," *Terrorism Focus*, Vol. 2, No. 7, March 30, 2005a. As of January 7, 2010:
http://www.jamestown.org/programs/gta/
single/?tx_ttnews[tt_news]=30194&tx_ttnews[backPid]=238&no_cache=1

———, "New al-Qaeda Outfit Announces Its Presence in Algeria," *Terrorism Focus*, Vol. 2, No. 10, May 31, 2005b. As of January 7, 2010:
http://www.jamestown.org/programs/gta/
single/?tx_ttnews[tt_news]=491&tx_ttnews[backPid]=238&no_cache=1

————, "Declining in Algeria, GSPC Enters International Theater," *Terrorism Focus*, Vol. 3, No. 1, January 9, 2006. As of January 7, 2010:
http://www.jamestown.org/programs/gta/
single/?tx_ttnews[tt_news]=634&tx_ttnews[backPid]=239&no_cache=1

UNMIK—*see* United Nations Interim Administration Mission in Kosovo.

United Nations High Commissioner for Refugees, "Statistics on Displaced Iraqis Around the World," September 2007. As of January 13, 2010:
http://www.unhcr.org/470387fc2.html

United Nations Interim Administration Mission in Kosovo, "Fact Sheet Kosovo," May 2003. As of January 7, 2010:
http://www.unmikonline.org/eu/index_fs.pdf

Urban, Mark L., *Big Boys' Rules: The Secret Struggle Against the IRA*, London: Faber, 1992.

U.S. Army Field Manual, *Counterinsurgency Operations FM 3-07.22*, Headquarters Department of the Army, United States Government Printing Office, 2006.

U.S. Census Bureau, *World Population Profile*, Washington, D.C., WP/96, July 1996. As of January 7, 2010:
http://www.census.gov/ipc/www/wp96.html

————, *International Data Base*, updated December 2009. As of January 7, 2010:
http://www.census.gov/ipc/www/idb/

U.S. Department of the Army, and U.S. Marine Corps, *Counterinsurgency*, Washington, D.C., U.S. Army field manual 3-24, Marine Corps warfighting publication 3-33.5, December 15, 2006. As of January 19, 2010:
http://www.fas.org/irp/doddir/army/fm3-24.pdf

————, *Counterinsurgency Operations*, Washington, D.C., U.S. Army field manual 3-24, Marine Corps warfighting publication 3-33.5, joint publication 3-24, October 5, 2009. As of January 6, 2010:
http://www.dtic.mil/doctrine/new_pubs/jp3_24.pdf

U.S. Department of Defense, *Terrorist Group Profiles*, Washington, D.C., 1988.

U.S. Department of State, *Patterns of Global Terrorism 2001*, Washington, D.C., May 2002. As of January 7, 2010:
http://www.state.gov/s/ct/rls/crt/2001/

————, "Nonproliferation, Antiterrorism, Demining, and Related Programs," February 13, 2006a. As of January 7, 2010:
http://www.state.gov/documents/organization/60647.pdf

————, "Sri Lanka," *Country Reports on Human Rights Practices 2005*, March 8, 2006b. As of October 17, 2006:
http://www.state.gov/g/drl/rls/hrrpt/2005/61711.htm

U.S. Government Accountability Office, *Plan Colombia: Drug Reduction Goals Were Not Fully Met, but Security Has Improved; U.S. Agencies Need More Detailed Plans for Reducing Assistance*, Washington, D.C., GAO-09-71, 2008. As of January 12, 2010:
http://purl.access.gpo.gov/GPO/LPS111445

U.S. Marine Corps, *Small Wars Manual*, Manhattan, Kan.: Military Affairs, Department of History, 1940.

Võ Nguyên Giáp, *People's War, People's Army: The Viet Cong Insurrection Manual for Underdeveloped Countries*, Honolulu, Hawaii: University Press of the Pacific, 1961 (2000).

Vreeland, James Raymond, "The Effect of Political Regime on Civil War: Unpacking Anocracy," *Journal of Conflict Resolution*, Vol. 52, No. 3, June 2008, pp. 401–425. As of March 25, 2009:
http://jcr.sagepub.com/cgi/reprint/52/3/401

Weinstein, Jeremy M., *Inside Rebellion: The Politics of Insurgent Violence*, New York: Cambridge University Press, 2007.

Westad, Odd Arne, "Concerning the Situation in 'A': New Russian Evidence on the Soviet Intervention in Afghanistan," *Cold War International History Project Bulletin*, Vol. 8–9, Winter 1996–1997, Section 2: US-Soviet Relations and Soviet Intervention in Afghanistan, pp. 128–132. As of January 7, 2010:
http://wilsoncenter.org/topics/pubs/ACF193.pdf

White House, "Lyndon B. Johnson," undated Web page. As of January 7, 2010:
http://www.whitehouse.gov/about/presidents/lyndonbjohnson

Wiest, Andrew A., *The Vietnam War, 1956–1975*, Oxford, UK: Osprey, 2002.

Wijesekera, Daya, "The Liberation Tigers of Tamil Eelam (LTTE): The Asian Mafia," *Low Intensity Conflict and Law Enforcement*, Vol. 2, No. 2, Autumn 1993.

Willis, Michael J., "Containing Radicalism Through the Political Process in North Africa," *Mediterranean Politics*, Vol. 2, No. 2, July 2006, pp. 137–150.

Wilson, Dick, *The Long March, 1935: The Epic of Chinese Communism's Survival*, New York: Viking Press, 1971.

Wilson, Scott, "Peru Fears Reemergence of Violent Rebels: Shining Path Movement Aided by Drug Traffickers, Police Allege," *Washington Post*, December 10, 2001, p. A18.

Wright, Thomas C., *Latin America in the Era of the Cuban Revolution*, New York: Praeger, 1991.

Yacoubian, Mona, *Algeria's Struggle for Democracy*, New York: Council on Foreign Relations, Studies Department occasional paper 3, 1997.

Yapa, Vijitha, "Tamil Lorry Bomb Rips Apart Central Bank in Colombo," *Independent* (UK), February 1, 1996.

Zartman, William, "Dynamics and Constraints in Negotiations in Internal Conflicts," in I. William Zartman, ed., *Elusive Peace: Negotiating an End to Civil Wars*, Washington, D.C.: Brookings Institution, 1995, pp. 3–30.

Zhai, Qiang, *China and the Vietnam Wars, 1950–1975*, Chapel Hill, N.C.: University of North Carolina Press, 2000.